DARWIN'S UNFINISHED BUSINESS

"Simon G. Powell forcefully but gently demonstrates that intelligence (modes of being that acquire information, learn, and meaningfully respond to larger contexts) is intrinsic to our natural world. People who deny the intelligence of the living (microbes, plants, other animals) are abysmally, indeed dangerously, ignorant. They literally ignore our ultimate, planetwide sources of joy, air, water, food, and energy. Read this informative, science-packed, yet accessible book and enjoy its wisdom."

LYNN MARGULIS,
AUTHOR OF *SYMBIOTIC PLANET*, MEMBER OF NATIONAL ACADEMY
OF SCIENCES, AND PROFESSOR AT UNIVERSITY OF MASSACHUSETTS-
AMHERST, DEPARTMENT OF GEOSCIENCES

"A treasure trove of interesting material."

EDWARD GOLDSMITH,
ECO-AUTHOR AND FOUNDER OF *THE ECOLOGIST*

DARWIN'S
UNFINISHED BUSINESS

The Self-Organizing
Intelligence of Nature

SIMON G. POWELL

Foreword by Dorion Sagan

Park Street Press

Rochester, Vermont • Toronto, Canada

Park Street Press
One Park Street
Rochester, Vermont 05767
www.ParkStPress.com

Text stock is SFI certified

Park Street Press is a division of Inner Traditions International

Library of Congress Cataloging-in-Publication Data

Powell, Simon G.
  Darwin's unfinished business : the self-organizing intelligence of nature / Simon
G. Powell ; foreword by Dorion Sagan.
      p. cm.
  Includes bibliographical references and index.
  ISBN 978-1-59477-440-9 (pbk.) — ISBN 978-1-59477-801-8 (e-book)
  1. Evolution (Biology)—Philosophy. 2. Darwin, Charles, 1809–1882—Influence.
3. Philosophy of nature. 4. Self-organizing systems. 5. Intellect—Philosophy. 6.
DNA. 7. Life. I. Title.
  QH360.5.P69 2012
  576.8'2—dc23
                                                                                    2011035985

Printed and bound in the United States by Lake Book Manufacturing
The text stock is SFI certified. The Sustainable Forestry Initiative® program
promotes sustainable forest management.

10 9 8 7 6 5 4 3 2 1

Text design by Virginia Scott Bowman and layout by Jack Nichols
This book was typeset in Garamond Premier Pro with Myriad Pro, Swiss 721 BT,
Trajan Pro, and Warnock Pro used as display typefaces

To send correspondence to the author of this book, mail a first-class letter to the
author c/o Inner Traditions • Bear & Company, One Park Street, Rochester, VT
05767, and we will forward the communication, or contact the author through his
website at **www.simongpowell.com**.

To my mother and father

I would like to thank Inner Traditions for continuing to support and publish my work. Thanks also to my furtive friend and collaborator Albert Catt and his assistant Morgan Russell. I am also indebted to Iain Lewis for many long and fruitful discussions concerning the core ideas outlined in this book. And many thanks must be given to Dorion Sagan, who encouraged me right from the start.

CONTENTS

FOREWORD

By Dorion Sagan

When we look back in time over the evolutionary landscape of the Earth, we find something surpassingly strange: the "primitive" and would-be antiquated beings of yesteryear outdo us in so many ways. Indeed they do so with such technical prowess that we may be forgiven the suspicion that their artistry is superior to ours not only in direct skill but also in both the dependable application and graceful conceal-ment of such skill. Thus, we precocious apes jabbering about the surface of the planet, as we commandeer electromagnetic means to broadcast pictures of our bullying and semi-clad bodies, appear like a species in the thrall of a global inferiority complex, trying to compensate by call-ing attention to ourselves. Don't believe the hype. We are not so smart. The brittle star species *Ophiocoma wendtii* may seem primitive, but this relative of the starfish eludes predators by means of a crystalline lattice of calcite integrated into its skeleton. A marvel of optical computing far beyond the designs of artificial vision labs, *O. wendtii*'s entire body acts as a compound eye focusing light and *seeing*.

Over three billion years ago bacteria evolved and linked ecosys-tems of sulfide oxidizers and sulfide reducers, thereby solving a prob-lem of living without polluting the environment. Bacteria also evolved ecologically balanced nitrogen fixation (compared to which human agribusiness's use of fertilizer looks like a bad joke), worldwide gene trading (a kind of combination of the Internet and genetic engineering to help evolution along), and the ability to grow a mile beneath the surface of the Earth in rock by means of making use of ambient chemical reactions.

The discipline known as DNA computing began when scientist Leonard Adleman solved a mathematical conundrum known as the "Traveling Salesman Problem" (which involves working out the shortest path that takes in all points on a given route). By generating DNA sequences representing all possible routes and mixing them in a test tube, Adleman was able to answer a problem that had eluded the most powerful supercomputers, whose few thousand processors pale next to the natural intelligence of DNA. Similarly, ciliates such as paramecia, which trade their nuclei, appear to have pioneered certain robust information management techniques to search and preserve connected data long before the advent of computer programmers; the DNA of these organisms can be a thousand times more per cell than for bacteria. It has even been theorized (by my mother, National Academy of Sciences member biologist Lynn Margulis) that our brains make use of tubulin protein-mediated neural processing systems once used for locomotion in anaerobic bacteria known as spirochetes. The notion of natural intelligence for which Simon G. Powell so eloquently argues in this volume must be taken as more than metaphor.

But if a close look at the evidence reveals that we cannot separate ourselves from other life-forms by our creative intelligence, so too would it be foolhardy to single ourselves out as agents of Zoroastrian evil, somehow endowed with supernatural powers of destruction. Some mushrooms, tempting insects to eat them, have genetically arranged to sprout through the organisms' bodies, using the insect head as a fungal flowerpot. As the frightening use of bioterrorism makes clear, bacteria can be powerful agents of destruction. Arthromitus bacteria, normally found in the hindguts of termites and wood-eating cockroaches, with a slight genetic change become spore-forming anthrax bacilli. The bacteria are not very adept at living in the bodies of mammals; if they were they wouldn't be so destructive of the hosts that support them, which is why they stay holed up in the ground, doing nothing for ages in the form of spores. Frightening as it is, their mode of propagation, causing sickness and mass hemorrhaging, and thus being returned to the ground in animals' blood, is not very efficient. Even with technological inducements they are far less virulent than smarter organisms that have learned to grow together symbiotically.

In light of these instances of life's innovation and creativity, what scientific justification is there for regarding ourselves as the smartest or most destructive or most conscious species? Prudence is inimical to the bluster and rampant growth we display as individuals, cultures, and a species. Persistent species may have wild youths, but eventually they must find a more stable place in Nature. Simon G. Powell's book may be a wake-up call showing us our true place in a Nature abundant in intelligence and sentience, if not consciousness.

DORION SAGAN is a science writer, essayist, theorist, and author of sixteen books, including *The Sciences of Avatar,* and coauthor of *Death and Sex* and *Into the Cool.*

"I speak about universal evolution and teleological evolution; because I think the process of evolution reflects the wisdom of nature. I see the need for wisdom to become operative. We need to try to put all of these things together in what I call an evolutionary philosophy of our time."

JONAS SALK

PROLOGUE

Darwin's Evolving Legacy

Over a hundred and seventy years ago an observant young man named Charles Darwin began to formulate a scientific theory that would dramatically change our view of life on Earth. Darwin's bold assertion that life has evolved over time through a gradual process of natural selection proved to be so controversial that its veracity is even now being fought over tooth and nail. Darwin verily rocked the world, and things have yet to settle down. Evolution is an issue that has had everyone from the Pope to Supreme Court judges earnestly debating its merits as a principle that can account for the existence of complex life. Despite the accumulation of evidence in favor of biological evolution (e.g., the existence of mutable DNA within all organisms, common anatomy, vestigial organs, the fossil record, etc.), people remain divided. Science avidly champions evolutionary theory, whilst many religious people avidly rally against it. Indeed, in the USA apparently half the population refuses to believe that the human race arose through an evolutionary process.

In the wake of these disputes, the intelligent design movement has sprung up. Based in the USA, the movement claims that evolution through natural selection cannot, by itself, explain certain intricate features of organisms. The inference is that some sort of omnipotent supernatural influence (i.e., an influence issuing from "outside" of Nature) is able to manipulate life and guide evolution. Even though most scientists accuse the movement of peddling an untestable brand of badly concealed creationism, the readiness with which intelligent design arguments are bandied about highlights the suspicion in many people's minds that conventional evolutionary theory is incomplete and "missing something."

Let me lay my own cards on the table. I submit that we are justifiably right to bark at the current orthodox view of evolution. However, the proponents of intelligent design are barking up the wrong tree. The problem lies not with evolutionary theory per se, but rather *the way in which evolution is perceived, appraised, and delineated.* The official line is that evolution can be defined as a change in a gene pool over time, or, to be more specific, a change in gene frequencies over time. Short and masterfully blunt, this kind of definition is found in most biology textbooks—yet it simply does not cut it. Such a definition is not wrong, but neither is it right enough. In a way, it is like defining the development of this book as a change in a text pool over time, or a change in word frequencies over time. This is true of course, but it is not true enough.

Given that the evolution of life is one of the most impressive and astonishing processes we know of, characterizing evolution merely as a change in gene frequencies over time is a lame definition that deserves barking at. Likewise, if polar bears, lizards, dragonflies, and the metabolizing cells of which they are built are considered as nothing more than the outcome of certain combinations of genes, then it is no wonder that many people are suspicious of Darwin's theory. Moreover, people are perfectly correct to infer some sort of intelligent design (albeit natural) throughout the tree of life. Not just in certain ultra-complex organs, but in any part of any organism. Even the most cursory glance at a biochemistry textbook highlights the mind-bogglingly impressive molecular machinery that, with precision and finesse, underlies all forms of life. Indeed, biological science is currently uncovering so many new and marvelous facts about genetic processes as to necessitate the creation of a whole new language. As exotic technical words are invented daily in order to label the newly discovered molecular machines and molecular routines being observed within cells, biochemistry is coming to sound more and more like the study of some highly advanced alien technology. How molecular biologists manage to keep a lid on all this is anyone's guess. And yet unfathomable supernatural forces are not required to explain these remarkable facts of life. Nature, by way of evolution through natural selection, is the intelligent design force, or intelligent system, that weaves together intelligently functioning biological structures.

The idea that Nature's evolutionary processes involve, and produce, intelligence may sound straightforward enough—yet it does not sit well with orthodox science or, oddly enough, with intelligent design enthusiasts. "Intelligence" is one of those words that we generally preserve for ourselves alone and not for something as seemingly abstract as the larger system of Nature. Hence, science flatly denies that biological evolution is intelligent, claiming instead in no uncertain terms that evolution (and, by implication, Nature) is blind, dumb, and utterly mindless. On the other hand, religious-minded scientists (like the proponents of intelligent design) insinuate that the intelligence involved with the existence of life lies mysteriously outside of Nature, rather that seeing intelligence as being a fundamental property of Nature or a fundamental aspect of living things. This book is therefore an attempt to venture beyond both the orthodox scientific view of evolution as well as the intelligent design creationist view, to gain a fresh perspective on evolution that might warrant acceptance from everyone. No new processes are invoked in order to achieve this new perspective. No supernatural phenomena are summoned. It is rather the case that life and evolution are *reinterpreted*—in the same way that one might reinterpret the meaning of a book or an ancient hieroglyph.

As intimated, my chief contention is that evolution through natural selection is a process that displays the chief characteristics of intelligence. I also contend that the laws of Nature, which facilitate evolution, are likewise bound up with intelligence. I portray this new perspective in terms of *natural intelligence* (thus making it distinguishable from, say, artificial intelligence or human intelligence). I do not mean that Nature is conscious, or that evolution has foresight, or that natural intelligence is exactly like human intelligence. What I mean is that intelligence is essentially an information-gaining capacity in which learning takes place, problems are solved, and sense is made—and that this kind of clever sense-making capacity is not restricted to the brain/mind. Nor is it restricted to robots and other examples of artificial intelligence. On the contrary, Nature, the Universe, can be seen as one vast system of self-organizing intelligence, biological evolution being one particular way in which this intelligence gradually but inexorably expresses itself. By the end of the book, I hope to have convinced the reader to take on board this seemingly audacious notion.

Taking intelligence in its normal usage, if you are intelligent you can learn about, and make sense of, the world around you (and, of course, make sense of an IQ test). The more sense you can make, the smarter you are likely to be. The more environmental information you can access, store, and organize, the more ingenious you can be in your behavior in that environment. Take a native tribe that has existed in the Amazon for millennia. Such a tribe evinces intelligence because its members have made good sense of their jungle environment and can thereby live and survive there. They have learned how the jungle environment works and can use this knowledge to their advantage. Or take a prominent historical scientist like Michael Faraday, immersed as he was in the academic environment of physics studies and chemistry studies: Faraday was highly intelligent because he managed to make sense of electromagnetism and paved the way for our electricity-using culture. (Note here that in both examples sense is/was being made of something that was always "out there" and already sensible. The realization that Nature *affords learning* is something that I will go into much more detail about later, as it bears heavily upon the process of sense-making.)

Now, the process of evolution likewise demonstrates the capacity to actively make sense. It achieves this through the vehicle of *bio-logic* (by "bio-logic" I mean the network of physiochemical processes found inside living things). Indeed, the kind of sensible biological engineering achieved by evolution ranks higher than any human technological achievement. Think of how difficult it is to build a biosphere populated by 100 million species—from scratch. Or how to overcome gravity and fly. Or how to photosynthesize (i.e., feed upon photons of light). Or how to build antifreeze agents. Or how to live in boiling sulfurous water. Through evolution, life has learned a plethora of ingenious sense-making techniques—such as specific protein sequencing, metabolism, motility, morphogenesis, cellular orchestration, aerobic and anaerobic respiration, self-repair, temperature regulation, replication, making use of electrochemical potentials, symbiosis, sight, hearing, the use of tensegrity (to achieve stable cell structure), and so on. These nifty techniques—all of which crucially allow living things to make sense of the world and thus be—are written down in enduring (but still mutable/ flexible) DNA. Indeed, genomes represent vast libraries of learning, a form of digital text scripted, honed, and refined over millions of years. The human genome, for instance, contains information pertaining to the production and orchestration of 100 trillion cells into a tightly bound organism potentially capable of warding off death in excess of one hundred years. A single fertilized human egg cell can gracefully differentiate into lungs, liver, spleen, heart, muscle, tendons, sinew, stomach, urinary tract, brain, nervous system, blood, bone, eyes, ears, tear ducts, skin, and so on—all knitted together into one cohesive self-organized whole.

Such ingenious biological behavior clearly evinces intelligence of some kind, as does the evolutionary process that wrote the human genome and scripted the behavior, and as do the laws of Nature that foster all this creative activity. Not necessarily conscious intelligence—Nature does not need to consciously "know" what it is doing through natural selection, nor do cells need to consciously "know" what they are doing (although awareness or sensation of some kind is not ruled out). It is rather the case that bio-logic behaves in a remarkably smart and sensible manner, as does the process of evolution that continually edits and refines bio-logic.

Evolution through natural selection is therefore much more than simply a change in gene frequencies over time. Evolution involves a cumulative series of *sensible genetic changes*. Sensible changes to genetic information equate to sensible biological outcomes. And sensible biological outcomes are, of course, those that ensure survival. In other words, changes in bio-logic that make some kind of good life-affirming sense in the light of Nature, or in the light of the environment, are the ones selectively favored by Nature. So while Darwin himself defined evolution as descent with modification, it is more accurate and more informative to speak of descent by way of *sensible* modifications to genetic information. Ipso facto, this gives evolution an inherent direction. Because there are specific ways in which to make sense (due to things like the specific laws of Nature and the specific manner in which atoms, molecules, and energy behave), this channels evolution into specific sense-making pathways. So where natural selection favors, say, a specific chemical cycle that yields useful energy, this means that bio-logic must perforce be attuned to the equally specific chemical and molecular laws that govern energetic chemical reactions. Indeed, it is precisely this inherent tendency for natural selection to engineer specific solutions to the art of living that explains the phenomenon of convergent evolution, where we see the same sensible organ or the same pattern of sensible biological behavior evolving in multiple branches of the tree of life. Thus, the irrepressible pop cultural meme "the survival of the fittest" faces immanent extinction, and a new and much more apt meme can take its place, namely *the survival of that which makes sense*. It is this sensible aspect of genetic change—the fact that Nature repeatedly favors the survival of genetic information that elicits sensible biological behavior—that characterizes the intelligence intrinsic to the process of evolution.

With current thinking, however, we tend to associate intelligence solely with human behavior and find it difficult to ascribe intelligence to physical, chemical, or biological processes. This is most likely due to the fact that our language is highly anthropocentric. It's no surprise then that the term "biotechnology" refers to our manipulations of bio-logic and not to bio-logic itself. So whereas we will ascribe intelligence to, and be impressed by, genetic engineering procedures that can make cows secrete insulin in their milk, we will not generally view mamma-lian lactation itself as being an impressively intelligent biological process.

Even if we do concede intelligence outside of our species, it is usu-ally only because of some sort of relatively simple tool use displayed by another creature. Since a tool is an object designed to perform a spe-cific useful function, we readily associate tools with intelligence. Thus, we might conclude that chimpanzees are intelligent to some degree if they can break open a nut by pounding on it with a carefully chosen stone. Or if a chimpanzee utilizes a carefully chosen twig to fish for termites, then we likewise infer that some degree of intelligence is being exhibited. But what of the chimpanzee's hands? Are they not tools for grasping objects? And what of its hand–eye coordination? Are the brain and nervous system of the chimpanzee not tools for making sense of the world around it? And what of its breathing as it carries out its nut-pounding endeavors? Are lungs not tools for imbibing oxygen? And are the molecules of hemoglobin in blood not tools for grasping individual molecules of oxygen so as to transport them around the body? Indeed, isn't the circulatory system an organismic tool for transporting nutri-ents, gases, and wastes? To be sure, if you were to carefully examine each and every part of the chimpanzee's body (or any organism), you would find system upon system of functional organelles, tissues, and organs—all of which are way more impressive than nut-cracking with a stone. The overt use of external tools rests upon, and is dependent upon, a huge hierarchical pyramid of internal biological tools. In other words, a genome is a code utilized to make vast arrays of sensible biological tools that are woven together into a single cohesive multifunctional totality. This totality—the organism—can then make tools outside of itself.

The life-enhancing biological tools that genomes code for have been designed by Nature through the process of evolution through natural selection—not necessarily consciously designed as we might consciously design something, but designed in the sense of having been shaped and honed to a high degree of specificity via the action of natural selection over millions of generations. Even the process of evolution itself can, if viewed over a long period of time, be seen as a tool for making tools. Once you begin to expand your concept of what a tool is, it becomes apparent that intelligence goes deeper than relatively simple overt behavior involving external tools (which we happen to be good at using). Given that everything behaves in a certain way (this includes bio-logic and all and any other part of Nature), then intelligent behavior can be extended beyond human behavior alone. *A process/behavior of any kind that gains information, learns, and actively makes sense of some larger context is, I submit, an intelligent process/behavior*—or at the very least such a process embodies the essential principles of intelligence to some degree.

My point, then, is that intelligence manifests as a particular kind of process not restricted to human behavior, and that if we are to debate the process of evolution with any kind of justice, then we must perforce admit that evolution operates intelligently in terms of the sensible patterns of behavior that it engineers via bio-logic. After all, it was evolution that crafted the human brain. To state that we are intelligent but that the process that sculpted our brains was not, seems a tad suspect. As does the notion that the vast systems of brain neurons underlying our minds lack any intelligent activity of their own. Human intelligence must, in some essential way, rely upon the brain, and the specific way in which the brain organizes information. These brain mechanisms are assuredly behaving and functioning in a clever manner (so clever, in fact, that science has a hard time understanding the full workings of a single neuron, let alone the workings of the entire brain). Similarly, if we imagine (courtesy of our clever brains) the long evolutionary line of ancestral hominid brains that led up to the modern human brain, it is clear that a progressive and ingenious engineering process is at work. Granted, this might have taken millions of years and involved much pruning at the hands of natural selection, yet the efficacy and sheer architectural prowess of Nature are, in this instance, undeniable. In my mind at least, the inference of natural intelligence within Nature is the only reasonable way of appraising both the evolutionary process and the complex organisms wrought by evolution. All the established facts about life do not change, only the *interpretation of those facts.* Hence, I propose that the concept, or paradigm, of natural intelligence be brought to the debating table and be given serious consideration by anyone interested in the meaning and significance of life.

All of which I speak is of the utmost importance. Indeed, paradigms concerning evolution and our place within Nature rank extremely high in the collective psyche. This explains why there is so much friction and discord between religion and science, and why there is a veritable industry of heated debate. Each group becomes more and more vociferous as the debate intensifies and the stakes escalate. A resolution must be forthcoming—for the simple reason that an objective truth exists. As I have said, I am convinced that the root of the discord lies with the way in which we conceive of evolution. We look at details and tend to lose sight of the bigger picture. Only when we look at the bigger picture and see Nature as a single coherent sensibly behaving system can we properly divine natural intelligence. If we spend too much time analyzing details, we will miss the larger context of which those details are a part. This is especially the case when the issue of genetic change is explored. By acknowledging that natural selection involves the preservation of sensible modifications to the genetic text and not just any old modifications, one is forced to consider the larger environment that provides something to make sense of. Because Nature is sensible *a priori* ("a priori" means "given before the fact") due to natural laws and the various effects of those laws, sense can be made of Nature by way of evolving bio-logic. This realization suggests that Nature can be viewed as a system of self-organizing intelligence in which life, consciousness, and information are all falling into place according to some unstoppable natural imperative.

Having said as much, it is clear that the paradigms currently steering human culture, particularly in the West, bear no reference whatsoever to natural intelligence. This might well explain why so many environmental and existential crises face us. Conventional science resolutely denies that any intelligence is involved in evolution, let alone other natural processes, and so it is principally human intelligence and all the inventions and ideas of human intelligence that are esteemed, and that come to determine the progress of human culture. In this sense we are going it alone, a species intent on divorcing itself from the natural intelligence that birthed us. It might well be that until we acknowledge the role of natural intelligence in the evolution of the tree of life we shall remain estranged from the rest of Nature and continue to suffer the consequences. In all fields, right relationships bring health and balance. To acknowledge the presence of natural intelligence within Nature, to seek to learn from that intelligence, is to begin placing oneself in a right relationship to all else; so poised, balance and health are maximized. And so we begin.

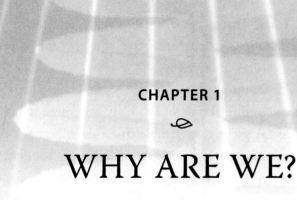

CHAPTER 1

WHY ARE WE?

As these words are written, a hefty space telescope is in orbit high above Earth. Launched back in 1990 and still going strong, the Hubble telescope has transmitted back to Earth a series of the most astonishing photographs of the Universe to date, photographs so provocative in their detail, so spectacular in what they purport to show, that one looks upon such images aghast and at a loss for superlatives. One of the most famous of these pictures shows an endless number of galaxies, each one an almost abstract splash of color. This particular shot resulted from Hubble focusing its reflecting lens upon an area of dark space equivalent in size to a grain of sand held at arm's length. And yet within such a minuscule area we see a plethora of cosmic structures—not just single stars but complete galaxies, each containing billions of stars and perhaps a multitude of solar systems like our own. The mind boggles. New York's sprawling metropolis appears big. The Grand Canyon is considered massive. China is vast. The entire biosphere is decidedly gargantuan. The solar system is so big that it takes eight minutes for the sun's light, traveling at more than 180,000 miles a second, to reach us. Yet the scale of the scenes captured by Hubble dwarfs our notions of largeness. In fact it is virtually—if not actually—impossible to fully grasp the kind of enormity associated with dozens of galaxies portrayed in a single image. Indeed, one could be forgiven for suspecting these pictures to be innovative fakes, created perhaps by some artful student on one of NASA's computers. Surely, one thinks, the Universe cannot really be that great . . .

If one dwells upon the sheer magnitude of the Universe, its seemingly limitless extent and fractal depth, then another more subtle wonder confronts us. Though it is incredible that the Universe is full of organized structures like stars, galaxies, and shimmering nebulae, more amazing still is our *conscious apperception of such a fact.* Somehow, and by some means, the physical and chemical laws of the Universe have conspired upon our planet to orchestrate zillions of atoms together in such a way that on a macroscopic scale fully functioning organisms result. And somehow, one such species of organism—we *Homo sapiens*—has been endowed with a perceptual capacity called consciousness, which can marvel at, and be amazed by, the very big Universe that managed to spawn it. Indeed, had such fine organic orchestration not come to pass and had consciousness not arisen, there would be no one to gasp in amazement at the epic majesty of the Universe—no "wow!" factor as it were. Above us lie billions of suns. Below us lies the same vastness. In fact, whichever direction we consider, we will find the immense depths of interstellar space dusted with suns. We can consciously reflect upon this, consciously know it to be true. Thus it is that our conscious minds lie suspended within a truly awesome cosmos.

Consciousness therefore lies at the heart of both our documentation and our appreciation of the Universe, for without consciousness, the Universe, as we know it, would simply not exist. Well, one might argue that the Universe "in itself" would still exist, but it would not be the Universe with which we are familiar, since the Universe we know is the Universe as focused through the unique perspective filter of the human mind. We cannot imagine the Universe without consciousness. It is only apparent to us through the medium of consciousness. In other words, consciousness is a fundamentally real property of the Universe, just as fundamentally real as the stars and galaxies revealed to us by the Hubble telescope. Attempting to conceive of a Universe devoid of consciousness is little more than a philosophical game indulged in by conscious minds. The very ground of our being is woven out of consciousness. Apart from blood and bone, we—*us*—are verily made of consciousness, and it is by dint of this fundamental characteristic that the existence of human consciousness within the Universe be just as extraordinary as the things perceived therein.

To really appreciate the fundamental quality of consciousness, however, one has to do a sort of double take. One must catch oneself or others in the act of being beguiled over some aspect of the cosmos and then strive to see that the persons themselves—their conscious apparatus if you like—are just as striking as anything else. No wonder then that some scientists refer to consciousness as the last great mystery. All our thinking, our philosophy, our ethics, our religion, our strife, our lofty ideals, our art, our science, our extensive knowledge about the nature and origins of the Universe—all this results because of the primary existence of consciousness. To paraphrase the eminent seventeenth-century French philosopher Descartes, we are conscious, therefore we think, marvel, love, hate, and muse.

A certain deep question inevitably comes to mind at this point, and this is the reason for the foregoing remarks. Sometimes it is only by framing the right question—seeing the validity of a question as it were—that we can thence seek out important answers and see things that are otherwise not so easily discernible. Actually, the question the reader is asked to consider is a question the aforementioned philosopher Descartes somehow managed to bungle even though he had an active interest in the Universe and the role of mind in the Universe. True, Descartes grasped the prime importance of consciousness with his dictum "I think, therefore I am," yet he failed to go on to ask the next obvious question. That question is "Why?" Why is it that consciousness exists? How and why did conscious minds like ours come to be? How is it that we can consciously build and launch orbiting telescopes and then consciously delight in the pictures taken? What does it mean for the Universe to have engendered conscious minds? Does this represent some jest? A trick? A strange plot perhaps? Or is mind merely the result of a "jammy" combination of physics and chemistry—a cosmic fluke in this particular neck of space? How much significance can one really attach to consciousness? In essence, is mind a curious lucky accident or was mind planned to emerge in some way?

THE HUMAN BRAIN

The first key point to make is that we are endowed with conscious-ness because we possess brains. Brains underlie minds just as ink and paper underlie books and computer hardware underlies the computer software and computer programs we use. And quite big and convoluted brains at that—human brains are apparently three times the size of chimpanzee brains (not that this necessarily makes us three times as smart as chimps, of course). This association of consciousness with the brain might seem very obvious, yet even Descartes did not dwell on it for too long (presumably because the neuroscientific study of the brain had yet to emerge at that point in history). As it happens, Descartes thought that the mind was somehow separate from the brain, and many people believe this even today (Descartes believed that the mind some-how interfaced with the body through the pineal gland). Here we have alighted upon the age-old dualistic mind/brain problem. And yet if you care to hit someone hard on the head, then his or her consciousness will doubtless "be lost" (this is what boxers try to achieve in the ring). Or if you infuse the neurons (nerve cells) of a human brain with an anaes-thetic compound—which reduces the electrochemical messages being transmitted between neurons—then once again consciousness seems to vanish (it should also be noted that patients who have had their pineal glands removed still claim to be conscious).

Perhaps a more profound indication of the intimate connection between the neuronal brain and consciousness lies with the operation of psychoactive drugs. Walk into any drinking establishment just before closing time, and it will be patently clear that those persons under the influence of alcohol are experiencing a modified state of consciousness. Similarly, a person under the effects of a psychoactive agent like LSD, cocaine, or Ecstasy will likewise be experiencing a dramatically altered state of consciousness. In each case, we see that chemical substances enter the brain, alter patterns of neuronal firing (which are mediated partially through chemical processes), and at the same time alter the quality and quantity of consciousness. This is even true of mild psychoactive substances like nicotine and caffeine. Most of us will know how a strong cup of tea or coffee in the morning can help wake us up. This is because caffeine is a chemical stimulant able to arouse the brain such that consciousness and attention become mildly enhanced.

Consciousness is thus intrinsically linked to the brain—all the evidence points to this, although in Descartes's day less was known about the brain and its functions. Scientists now think the brain processes information in an analogous manner to the way in which a computer processes information—it's just that the brain happens to be a biomolecular computing device far more powerful than any man-made computer.

So far so good then. Our deep question about the Universe and about consciousness can now be restated in the following way: Where on earth did the human brain—with its astonishing capacity to substantiate consciousness—come from? How did the Universe manage to give rise to consciousness-embodying brains? If the human brain represents the most complex information-processing device we know of, and if it is literally made of commonly available "bits" of the Universe (such as carbon atoms, for instance), then by what means did it arise? And why? For it surely did not spring out of some large fertile hat or cosmic cereal box for no reason. So this is still a good question and, as the reader will agree, an important one if we are to fully make sense of our place in the Universe at large. Certainly it would be missing the point to live and die without having thought about how we and our conscious brains got here.

THE EVOLUTIONARY IMPERATIVE

The most immediate answer to our progressive question is that the human brain exists because it *evolved* into existence. At some stage, evolution—the repeated action of natural selection—led to mammalian brains and the gradual complexification of mammalian brains. In other words, over millions and millions of years, mammalian nervous systems have evolved to the point in the hominid line where complex consciousness has emerged (or has been facilitated). Although science might not yet know each and every way in which a living cell monitors and regulates its own DNA, and although science might also be ignorant of certain subtle self-organizing properties of Nature that might facilitate biological organization in some way, it is certainly true beyond any reasonable doubt that evolution has happened, that life began relatively simply and gained in complexity over time, that all forms of life are intrinsically related, and, moreover, that evolution is afforded by means and mechanisms inherent within Nature and not by something supernatural.

Understandably, the public is still digesting the theory of evolution and its various implications. And there would appear to be major signs of indigestion. As I stated in the prologue, half of the world's most prosperous and most modernized nation refutes evolution. Apart from adherence to religious dogma, another reason that people are suspicious of evolution is doubtless due to a popular misconception. Many otherwise well-educated individuals mistakenly believe that evolution consists solely of chance, that evolution is synonymous with randomness, akin to a sort of lottery. Indeed, it is quite astonishing how often this misconception appears in debates about evolution. In such cases of marked misunderstanding, it is not appreciated that there are two principle components to evolution; namely variation/mutation, which is indeed assumed to be random, and natural selection, which is patently *not* random since the environment will invariably select precisely those variations/mutations that are more adaptive and that confer a reproductive advantage.

According to the above reasoning, then, evolution appears to be the answer to our question about why brains and consciousness exist. This being so, we might even agree with those who see Darwin's theory of evolution as being the single most important scientific concept of our time. So can we stop asking questions about the Universe now? Does the buck of explanation stop at evolution? Has the wonder and mystery of our conscious existence within the Universe been vanquished by this line of careful scrutiny? Can the single word "evolution" put our inquisitive minds to rest?

Most biological scientists will answer in the affirmative. They will invoke evolution to explain the existence of complex life and complex brains and then simply brush their hands as if no more explanation were needed, as if the tree of life were the simplest thing in the world to explain. Evolution is touted as a *done deal,* and we are encouraged to move on to some other phenomenon to ponder over. But these biologists do evolution, and Nature, a major disservice. Indeed, our big question has still to reach a genuinely satisfactory answer. For, feeling a trifle sublime, we might employ our brains next to ask: Okay, if evolution through natural selection has engineered conscious human brains, then *from whence originated this extraordinary creative potential?* If evolution accounts for the origin of species and the complex biological organs of which they are made, then what is the origin of this origin? How, and why, does Nature have the ability to repeatedly select certain configurations of DNA? Indeed, what exactly is happening when a particular genetic change is favored by Nature? What does it really mean when we say that Nature, or the environment, *selects* something? And if DNA is a language that underlies bio-logic, then why is Nature endowed with this language as well as the capacity to express it so eloquently and in such a flexible way? Come to think of it, why was the famous genetic code always waiting to fall into place? How on earth is all this remarkable orchestration possible in an apparently dumb and mindless Universe?

These deeper questions about the nature of evolution are probably the least asked by those scientists in the field presumably because such questions, despite their profundity, are deemed to be of no importance to evolutionary theory itself. Regardless of why Nature supports biological evolution, and regardless of what precisely is happening when Nature commits a selective act, the theory of evolution can account for all the varieties of life, and that is all that really matters. According to dogma, it's all just a simple matter of genes changing over time. Period. Or so they say.

Apart from shedding light on the sense and significance of human existence, the principal reason to ask deeper questions about evolution lies with the fact that evolution is the most astonishingly creative and constructive process we know of. I defy anyone to counter this assertion. No evolution and no DNA-writ life. No life and no organisms. No organisms and no complex biomolecular devices known as brains. No brains and no equally complex consciousness. No consciousness and no us: a Universe devoid of consciousness. It cannot be truly imagined.

Nature's epic process of evolution is thus responsible for our bodily existence as well as our experience of such a fact. And, for that matter, for the colorful wealth of biodiversity covering the globe. The rose in bloom, the song of the nightingale, the shower of pink cherry blossoms in spring, the elegance of an eagle in flight, the taste of strawberries and cream, even the gaseous composition of the atmosphere—we must acknowledge that the evolutionary process yielded the whole organically molded caboodle. Willy-nilly, evolution built everything that really matters. To state nonchalantly that evolution *just happens* and that it involves no more than changes in a gene pool over time, or that it is simply descent with modification, is really not good enough. A better and deeper understanding of evolution *demands* to be brought to the table. Nature is crying out for a more decent appraisal.

THE MEANING OF LIFE

In questioning the fundamental nature of evolution, we are really asking about the fundamental nature of reality, about why certain capacities and potentials are written into the fabric of Nature. And the ways in which we comprehend the nature of reality—so-called paradigms—are important since they can define our existence and can influence cultural drives and cultural value systems. Indeed, it can be said that the primary driving force of a culture resides in fundamental patterns of belief, much of which will be taken for granted. How we conceive of evolution will thus condition how we judge the meaning of life. And how we judge the meaning of life will condition how we treat life and subsequently live life.

The paradigms that underlie our conceptual vision of reality and our understanding of Nature are always constructed from axioms and assumptions. Such assumptions, such givens, are usually tacit, which is to say that they are rarely brought into the light of conscious awareness. They are just there, implanted within our minds from a young and impressionable age and never questioned, since to question those assumptions is akin to questioning the assumptions behind mathematics or the assumptions that paper money has real intrinsic value. However, if such tacit assumptions about reality are questioned and found to be in error, or at least found to be wanting in some way, then a paradigm shift might well ensue. When this happens, an entire conceptual framework of human thought and human thinking can be overwritten with a new set of concepts and underlying assumptions. We should pursue this realm of thought, since, as we shall see, our paradigm about what evolution is and what evolution represents rests upon a questionable dogmatic perspective.

As a way of exploring this issue of unstated assumptions guiding our conceptions about various things, think of the game of chess. The standard game expresses the underlying standard chess paradigm. All games of chess, from the simplistic to the masterly, unfold according to the initial axiom, which determines the precise configuration of all the pieces at the very start of the game. The rules of chess, or unquestioned background assumptions, then determine the moves each chess piece can make (and their powers) and thus allow the game to discretely proceed. The chess paradigm is just accepted, for to question it would be considered, among other things, rude. Certainly a Grandmaster would be thrown out of a chess tournament if he commenced a match by vociferously disputing the very axiomatic foundations of the game.

If, however, we do question the underlying framework of chess and proceed to restructure the game, then this would represent a paradigm shift in the "chess universe" since all games would now be different—the playing field and possible moves will have been radically altered at a fundamental level. One could even come up with a new piece and that, too, would significantly alter the paradigmatic nature of the game (imagine, say, two Trickster pieces, shaped like Jokers, and able to move differently according to which column or row each is placed on at the time).

A more classic example of a paradigm shift is the one instigated by Copernicus and Galileo. Until the Copernican revolution, learned people believed the sun to orbit the (static) Earth. Judging by the sun's daily course, it seems obvious to the immediate senses that the sun turns around the Earth. But by switching perspectives and realizing that in actuality the Earth orbits the sun, a rather spectacular paradigm shift was unleashed, one that caused humankind to seriously rethink its place within the Universe. It is interesting to imagine what this paradigm shift, or *metanoia,* must have felt like at the time. For many people, it must have felt as if one of their most cherished beliefs had been ungraciously snatched away from them. For others, the paradigmatic switch may have felt liberating.

This sort of dramatic conceptual reboot also transpired when Darwin introduced the theory of evolution. Darwin, a free thinker bravely unshackling himself from the religious dogma of his day, was questioning things—the rules and assumptions of belief if you like—that until that time had been taken for granted. The prevailing notion that a personal and omnipotent God (albeit Copernican) had created all and sundry in a single solitary act of creation and that species were immutable was tacitly assumed to be true. Such a creed was unquestioned by most people. It was accepted much as we now accept that the Earth goes around the sun and that the rules of chess are written in immutable stone. Darwin turned our normal frames of reference on their head and instigated a paradigm shift whose quakes and aftershocks are still with us. New perspectives can appear threatening, especially to those who cling hard to old and traditional ways of thinking. The main lesson to be learned from paradigm shifts is that we *interpret* Nature, and that our interpretations, as with interpretations of foreign texts, for example, can be more or less accurate. An erroneous interpretation is akin to an erroneous paradigm.

INTERPRETING EVOLUTION

Our current interpretations of evolution may likewise be in error—not in terms of the chief modus operandi of evolution as Darwin originally glimpsed it, but in terms of what evolution, as a process, really represents. If so, then a paradigm shift may be in the cards. What is of most importance here is whether *intelligence* is bound up with evolutionary processes. Does the reiterative context/fit aspect of natural selection embody the characteristics of intelligence? Indeed, can evolution be interpreted as an intelligent process yielding intelligently organized and intelligently behaving organisms? And, by extension, is Nature itself, in terms of its laws and its creative potential, an intelligently configured system whose function is to foster and nourish life? In short, is intelligence an essential aspect of the specific way in which the Universe's informational content flows, moves, and organizes itself?

Although these might seem obvious truths to some, the fact remains that current evolutionary theory does not view evolution as a process bearing the hallmarks of intelligence. Indeed, the very opposite holds true, namely that evolution is considered a wholly dumb process (i.e., no more than mindless and directionless changes in gene pools and gene frequencies) or at least as a process devoid of any quality such as intelligence. I surmise that this is primarily because of the limitations of our definition of intelligence. And yet, if an intelligent process is one that involves gaining information, learning, and making sense, then evolution through natural selection can be seen as an instance of intelligence in action. How so? Well, evolution certainly involves a gain of information—this occurs through the non-random accumulation of implicitly meaningful arrangements of DNA. That learning has taken place is demonstrated by the fact that life has, over billions of years, solved all of the major problems associated with living and being. As to sense-making, bio-logic is the art of sense-making made flesh. Literally. None of us would be alive were it not for the fact that our constituent bio-logic is able to constantly make sense of the environment. Hence the human organism can maintain a core temperature, can absorb food and manipulate energy, can excrete toxins, can synthesize hundreds of specific neurotransmitters, enzymes, and hormones, and so on—the examples of sensible biological behavior are endless. (Although we will be returning to this subject in much more detail in later chapters, the reader should now have the gist of what I mean when I talk about intelligence.)

We should also bear in mind here that an intelligent process does not necessarily need to be conscious. Robots endowed with computer systems can potentially learn about and make sense of their environment. And if a robot can indeed make good sense of its environment and can learn how to behave judiciously in that environment, then it is duly accorded some measure of artificial intelligence—even though, of course, such intelligence is not likely to be conscious. So intelligence does not need to be conscious (one can instead think of unconscious intelligence or even subconscious intelligence). It is rather the case that intelligence involves the capacity to learn and to actively make sense—and this is exactly what, at heart, biological evolution is all about. Life is good at making sense of its environment and that is why life exists and persists.

That our current interpretations of evolution fall short of the mark has been succinctly stated by, among others, noted ecologist Edward Goldsmith:

It is ironic that to explain what are the paltry, not to mention socially and ecologically destructive achievements of scientific and technological man, such as the invention of the internal combustion engine and the atom bomb, we invoke his consciousness, his creativity, and his intelligence, yet we categorically deny these qualities to all other living things, let alone to the miraculous processes of evolution and morphogenesis that brought them and him into being.[1]

Building upon this wholly warranted criticism that mankind has given Nature short shrift, the following chapters should convince the reader that our current interpretations of evolution are indeed flawed and inadequate—that when we look closely at what evolution is really all about then we shall see that, at heart, it is most assuredly an intelligent process whose essential property is literally to make, or construct, sense (i.e., make smart sensible structures) and that, *by logical necessity,* eventually had to elicit consciousness in some form or another.

This is indeed a radical interpretation of evolution, but it can be confidently stated at the outset that all the evidence supports it. Indeed, the more closely one looks at Nature and the evolutionary products of Nature, the more easily can one discern the intelligence operating therein. This intelligent or skillful quality of Nature as evinced by evolution can be termed *natural intelligence,* for that is what it is. It is precisely because of natural intelligence that the refined four-bit digital DNA code has arisen, that organic "machine code" that underlies all forms of life from bacteria to buffalo to bugs. Likewise it is because of natural intelligence that this DNA-based programming language has eventually endowed us hominids with brains blessed with consciousness and the attendant capacity to enthuse over the enormity and expansiveness of the Universe. Natural intelligence is therefore the answer to the question posed earlier, for natural intelligence adequately explains the emergence of evolutionary processes, the complex organisms so produced, the continual refinement of organisms, as well the subsequent emergence of human consciousness.

But before we begin to pursue this new approach to understanding evolution, we must first come to terms with the "old" paradigm, that paradigm currently taught both explicitly and implicitly in all our schools, colleges, and universities. In a way, this entails a step back from what has been said thus far—but only inasmuch as a step back is required before making a leap forward. In the next couple of chapters, then, I will be presenting the old dogmatic way of delineating evolution. In this way, we will be playing devil's advocate to the new natural intelligence paradigm, or at least setting up the scene whereby natural intelligence can be reintroduced beyond and above what has been said about it so far.

As stated, the commonly harbored paradigm of evolution has it that the emergence and continuous refinement of bio-logic is deemed not to be in any way reflective of intelligence. To be sure, the implication is that Nature is utterly mindless and purposeless and that evolution—while it may yield complex and wonderfully intricate living things—is nonetheless devoid of all and any intelligent characteristics. This entails a Universe that is purely accidental, indifferent, a sort of pointless freak incident that just happens to foster evolutionary processes. According to such a "whoops!" paradigm, as some call it, life really is accidental, hard, and tipped with death and nothing more. The evolution of life and consciousness emerges as a kind of gloriously daft ruse, an apathetic process going nowhere and doomed to eventually flicker out. If, however, we grant that our current understanding of evolution is incomplete, then our entire perspective on life has room to shift and change. Further, to view biological evolution as a naturally intelligent process is to imbue the biosphere with something more than simple wonder. The tree of life, complete with conscious nervous systems, emerges as a kind of monumental organic mirror testifying to, and reflecting, the fabulous intelligence inherent within the fabric of Nature. As the next few chapters serve to elucidate, when the evolutionary process is looked at in proper perspective and in proper context, a conclusion along these lines is inevitable. Any other view that avoids the invocation of natural intelligence simply will not do.

CHAPTER 2

∽

LIFE

A Great Organization

Despite its prodigious creative power, evolution is a process that has been seriously downplayed by modern science. The heart of the problem is that science has a tendency to be "merelyistic" in its approach to describing and delineating the evolutionary process and its fruits. Throughout the literature emanating from biologists and evolutionary scientists we find rampant expressions of "merelyism," in which the various intricacies and subtle organic gestures of life are invariably reduced to "merely this" or "simply that." The creed of merelyism abounds in popular books about evolution. Don't be fooled by the apparent design finesse exhibited by organisms, warn evolutionary biologists. It's not really design, for evolution is as blind as the bats that it generates. Evolution knows not what it does. It has no direction, no purpose, no point; in short, the evolving biosphere is a phenomenon completely devoid of intelligence (unlike us, of course). The evolution of the tree of life must, at all costs, be reduced in its grandeur to "mere blind groping." Well, not even groping really, because "groping" suggests some small semblance of awareness or intelligence.

Thus it is that we find evolutionary science going to the most extraordinary lengths in its attempt to break evolution up into mind-less components. Minuscule and inert parts of the evolving tree of life are proffered as evidence of the dumb and purposeless nature of evolution. It is as if the majority of evolutionary biologists are somehow a bit unnerved by the fact that Nature weaves such organic wonders as thermally insulated penguins, web-spinning spiders, and photosynthesizing plants. Again and again we are told that it is incumbent upon us to be not quite so astonished by the sophisticated design specifications of living things. Show a scientist an albatross able to soar effortlessly through the air and accurately navigate across vast oceans and he or she will invariably reduce the bird to the end point of a very long sequence of non-intelligent events, namely the gradual accumulation of directionless genetic changes. Move underwater and point to a gracefully swimming octopus with shade-changing skin, and the creature is likewise considered to be the product of a long series of blind evolutionary steps. The same blithe sentiments crop up when delineating the exquisite cellular activity transpiring within an albatross or an octopus. Although it is beyond the capacity of the most powerful and sophisticated computer in the world to model the life-affirming protein manipulation occurring in one single cell over the course of but a few nanoseconds, science steadfastly refuses to ascribe any kind of intelligence to the behavior of bio-logic.

Given the truly spectacular legacies of biological orchestration evinced in the tree of life, the merelyistic paradigm currently all the rage among biologists is a distinctly regrettable affair since it ultimately serves to belittle evolution, making light of an otherwise splendid and awe-inspiring process. In the following chapters a rather dramatic shift in paradigmatic perspective will be outlined according to which the evolutionary process can be viewed in a finer and more enhanced light. Moreover, this new paradigm can be delivered without recourse to supernatural forces. The paradigm of natural intelligence stands on its own, needing little more than common sense and a fresh outlook.

Actually, before we begin on this course, the basic psychological principle underlying the paradigm shift can be detailed. For the proposed switch in perspective is mirrored by the switch in perspective involved in viewing that famous face/vase picture. Most people are familiar with this image. It is the picture wherein the simple silhouettes of two faces gaze at one another. Yet a shift in perceptual viewpoint reveals a vase instead of two faces. Which image one sees depends upon the mind's eye. Neither perspective is false. Both are true, yet one view must dominate, since it is hard to see both images at the same time. With practice, one can readily switch between the two interpretations.

Conceiving of natural intelligence is analogous to conceiving one aspect of the vase/face picture. In this case, to switch conceptual perspective such that natural intelligence becomes apparent within evolution is not to deny the old dogmatic view of evolution but to reinterpret the evidence. Both paradigms hold true, just as the two paradigms in the vase/face picture hold true. Neither perspective is false. My claim, however, is that it is high time we took stock of the new view, a view that entails a reinterpretation of the process of evolution and can place us in a better psychological relationship with the biosphere and with the rest of Nature. Indeed, if we can divine natural intelligence, that is, if we can see that evolution is, ultimately, an intelligent process instigated by an intelligently configured Nature, then the corollaries are enormous and diverse. For one thing, the tree of life can be more appreciated, acquiring a more significant and valuable status. This is important when you consider that how we view life determines how we live life.

The invocation of natural intelligence is obviously in stark contrast to the merelyism currently fashionable within evolutionary science, that paradigm that seeks to explain life and evolution by appealing to life's smallest parts—like genes, for instance. Again, it is not that this merelyism is wrong, only that it is not right enough. Like the vase/face picture, it is but one part of the truth, a part that, in this case, comes into focus when looking at details. And details, of course, can blind one to higher-level meanings and higher-level patterns. Indeed, many of us might be unsettled by evolutionary dogma with its emphasis upon essentially static bits and pieces. This might not be because the dogma is flawed but because the dogma fails to capture the full brilliance of evolution, which is more apparent when evolution is viewed under more holistic magnification. As stated, merelyism is in danger of explaining life, and even consciousness, *away,* reducing everything of apparent significance to merely this or merely that mechanism. The three-and-a-half-billion-year-old tree of life surely deserves greater respect and admiration.

The emotional and intellectual failing of evolutionary dogma probably goes a long way in explaining why an alarming number of people still refuse to accept the theory of evolution. It is hoped that the paradigm of natural intelligence posited in these pages goes some way in addressing this important issue. Having said that, the fact remains that modern evolutionary dogma has many powerful adherents and any change in perspective will be, by necessity, hard won. In the following pages I make that hard fight using not wishful thinking but reason, logic, and above all else a basic common sense that no longer seems so common.

DISORGANIZING A HUMAN

Thought experiments come easy. To get people to think of them, the only equipment you need is imagination. You don't need to work in a lab or build any kind of elaborate recording equipment. Nor do you need to employ a host of collaborating experimenters, data collectors and statisticians. Despite their simple and inexpensive status, thought experiments nonetheless allow us to think in new ways and thus advance our understanding of the Universe. This, of course, explains why Einstein was so fond of using them. For example, Einstein used to imagine occupied elevators traveling at the speed of light. Apparently this kind of musing helped him develop his celebrated theory of relativity, a theory so profound and strange that it must be true.

As another more contemporary example, thought experimentation is utilized as a way to best approach the problem of contacting any hypothetical extraterrestrial intelligence. Team members of SETI—the NASA-based Search for ET Intelligence—undoubtedly spend a lot of their time in thought experimentation, for how else could they go about searching for signs of ET intelligence? Creative imagination is the key. Once we are primed to think in certain ways we can then reach conclusions and approach problems from new angles. Of course, this does not obviate the need for tangible physical experiments in which "real" data is gathered; thought experimentation is simply a useful tool to be implemented on the road to understanding. In a real way, thought experiments work in the same manner as metaphors. By directing our imagination, they can make certain things, or certain principles, very clear. Hence the strange subheading "Disorganizing a Human," which refers to a particular thought experiment useful to us at this juncture.

Before proceeding, however, recall that at this point we are principally interested in getting a handle on what evolution is all about, and what the existence of such an eminently creative process signifies. Since evolution is bound up with life and living things (and consciousness too, of course), then the following scenario allows us to cut straight to the chase. If we can establish what exactly living things are, we shall be in a position to contemplate what exactly it is that evolution does.

So let's take an example of life. Well, not any form of life, but a human being. In purely physical terms, a human consists of a finite amount of physical material, more or less, depending on whether the human in question is big or small. Of what does this physical material consist? Of what is a human made? Clearly, a human is made of flesh. And what is flesh? Flesh consists mainly of cellular matter, that is, matter organized in the form of individual cells packed en masse into a seamless organism: billions upon billions of cells tightly interlaced into a single living system. And what is a cell? A cell is one of life's most primary units. In its design and functionality any cell of a human being is like a cell inside a fish or a cell inside a plant. All organisms are made of cells. Scrape a leaf and the green residue will consist of cells. Put a bacterium under a microscope and one will see that it is a single-celled creature. Life is woven from cells.

But what are cells themselves made of? Well, if you really get down to it, cells are constructed out of huge conglomerations of molecules, some of which, inside the nucleus of the cell, compose DNA, the coded instructions that help determine the life, morphology, and activity of the cell. Thus, to say that a human is a molecular phenomenon is true enough at a certain level of analysis. And this molecular foundation of cellular matter is itself composed of even smaller bits. These are the atomic elements of which molecules themselves are made, basic atoms like carbon, hydrogen, and oxygen. So, at an even deeper level of conceptual magnification, a human being turns out to be an atomically constructed being. String atoms together and you get molecules. Arrange enough molecules together and you get cells. Amass cells together and you get organs and tissue. Fit organs and tissue together and multicellular organisms like humans emerge.

Now, let's take the conceptual liberty of liquidizing a human in a huge liquidizer. This might seem like harsh treatment, but sometimes one has to be cruel to be kind when it comes to delineating certain issues. All readers with a sensitive disposition should leave now. However, if you are willing to venture where few minds have gone before, then let us proceed with the human liquidization. First we get our subject to sign a disclaimer. Next, the subject places him- or herself in a large liquidizer. Then we don safety goggles and an apron, turn the big machine on, and stand well back . . .

Peering sheepishly into the liquidizer, we realize that we still have the same amount of matter, only now it has been rearranged in the form of an unsightly sludge of reddish porridge. There might be some undamaged cells in there, but our human subject has certainly gone. Well, to be sure, the subject is still there, but obviously not in the same form. Liquidize this human-sludge a little more on a higher setting and again we have a new configuration of elemental constituents and again a fully functioning human being is absolutely nowhere to be seen.

What we have done is taken our human subject apart bit by bit and then let all the bits and parts sit there in a slushy mess. The point of imagining this gruesome act is to allow us to start contemplating the nature of *organization;* to realize how different the matter in a fully working human is from that same matter redistributed into a featureless and disorganized muddle.

We can be less gory in our experiment. Instead of liquidizing our human, we could get some machine to partition him or her into a massively tall stack of cells. Or, more drastically, we could get a really sophisticated device, a reduction machine as it were, to decompose all of a human's cells into their constituent parts. This might result in a few large buckets of water, piles of carbon, calcium, and phosphorous, and a jar of nitrogen (doubtless this would make a neat art exhibition to rival those of maverick UK artist Damien Hirst).

Quite simple really, yet this is exactly what, at heart, any human is made of. Breaking the water into hydrogen and oxygen leaves us with six basic elements. Who could look at such a simple collection of matter and conceive that it could all be rearranged into a human being? No one. If you opened a traveling sideshow with these discrete materials and offered prize money to any bystander who could guess their origin, you would not have to worry about parting with your money. With even more abstraction, we could subdivide these six basic elements into trillions and trillions of atoms. Of all the possible ways in which these myriad atoms could be organized, there is basically only one sure-fire way in which to join each and every one of them into a healthy living breathing functioning whole. Not only is this way of reorganizing the matter unbelievably complex, it is also unique.

In this sense, then, *living organisms are highly unlikely combinations of matter.* Absurdly unlikely, in fact. The statistical chance that billions of atoms could form together into a living organism is beyond comprehension. It is akin to shuffling a million packs of cards and then expecting all of those cards to be arranged into perfect consecutive suits. If you pulled it off one afternoon no one would believe you. Or it would be like randomly spraying a billion tiny droplets of paint onto a wall and coming up with a perfect high-resolution picture of the Beatles' *Sgt. Pepper's* album cover.

In other words, organisms could not have arisen by way of chancy and disorganizational processes alone since, statistically speaking, there are an almost infinite amount of "useless" ways in which the atoms composing an organism could be arranged. This explains what is known as *entropy*. Entropy is a statistically determined property of a system pertaining to disorder or the lack of pattern. The main thing to realize is that there are more ways for a system to be *dis*organized than there are ways for it to be organized. Thus, if a system has high entropy it is said to be disordered and devoid of any pattern or any useful concentrations of energy. It is also of great import that entropy increases over time. This is why any highly ordered system will have a tendency to run down and "lose shape," as it were. Entropy also dictates why any random mixing of the matter of which we are made will lead us further and further away from our initially organized state. Entropy implies the move toward disorder and stale equilibrium, states of a system in which its components and its energy are spread equally about, and no patterns are evident (if this book were to succumb to entropy, for example, it would simply disintegrate into illegible fragments). The principle of entropy needs to be grasped since it appears, on the face of it, to be totally contrary to what evolution is all about. By understanding the statistical principle of entropy, we will see more clearly the significance of evolution.

GETTING A GRIP ON ENTROPY

The simplest example usually given to highlight entropy is to think of two boxes with a partition dividing them. If we have a concentration of hydrogen in one box and a concentration of nitrogen in the other, then we can say that the gaseous state of the two boxes corresponds to a far-from-equilibrium state, an unusually ordered state if you like. But open the partition and, through random molecular motion, the gases will diffuse so that eventually the system will settle down into molecular equilibrium, a state wherein the hydrogen and nitrogen atoms are distributed equally and evenly throughout the two boxes. This state corresponds to a maximum state of entropy or disorder. Everything is spread out as much as possible. The chance that, via random molecular vibration alone, the two gases would separate and somehow work themselves back into the two boxes is basically nil; it is a statistically improbable event. It would be somewhat like the spontaneous reformation of a human from a liquidized and disordered state—so unlikely as to be, barring miracles, impossible.

The same could be said of a sugar cube tossed into a mug of hot coffee. Its precise macroscopic-patterned structure will dissolve as entropy wields its statistical might and ensures that the sugar molecules will never spontaneously crystallize so as to re-form the original cube. According to the statistical rules of entropy, order has a tendency to dissolve and diminish. In simple terms, entropy is a bit of a cad since it tends to do away with order; it dissolves patterns, disintegrates things, reduces energy gradients, and pushes systems toward dull and dim states of equilibrium in which nothing interesting happens.

Organisms, then, represent extremely unlikely organizations of matter, intricate and complex patterns of basic elements that, together, manage to retain a coherent patterned structure. Whilst nonliving matter inevitably seeks equilibrium in which nothing interesting happens, living systems represent "islands of order," or decidedly complex patterns, within a sea of entropic disorder. As systems theorist Fritjof Capra remarks:

> The study of pattern is crucial to the understanding of living systems because systematic properties . . . arise from a configuration of ordered relationships. Systematic properties are properties of a pattern. What is destroyed when a living organism is dissected is its pattern. The components are still there, but the configuration of relationships between them—the pattern—is destroyed, and thus the organism dies.[1]

Life is better understood at this point as a *process* rather than a thing, a *verb* as opposed to a noun—moreover, a highly cunning verb whereby, as we have seen, entropy is locally circumvented such that statistically improbable patterns *are* created and *are* maintained. Sometimes the evolution of life is referred to as being *negentropic* since it is seemingly a contrary process to entropy. This is rather misleading, however, as entropy provides a *sensible flow* to energy and it is precisely this sensible energy flow, or current, that life has tapped in to in order to run itself. So just as a watermill can tap in to the entropic energy current of a waterfall and channel that downward-flowing energy into doing something useful, so too does life tap into the same kind of downward-flowing energy (i.e., entropy) and use it for its own ends. Metabolism is a good example of this. If we imagine that certain molecules are like coiled-up springs full of potential energy, entropy ensures that such springs will have a tendency to unwind and so release energy. This is what happens with metabolism. Through the inexorable flow of entropy (the spreading out of energy), chemical energy is released from food (springs are uncoiled) and harnessed to fuel life-affirming processes. Which means, ironically enough, that life uses the tide of entropy in order to swim against it, or at least to remain in the same place. When an individual living organism dies, however, entropy duly resumes in earnest and the constituent components of the deceased organism dissipate. The unique pattern, the exquisite organizational complexity, literally disintegrates, as it did when we so callously liquidized the subject of our thought experiment.

LIFE AND ENTROPY

Given the natural tendency for increasing entropy in the Universe, that is, given the statistically determined tendency for physical systems to run down and reach dull states of equilibrium, it is truly astonishing that, conditions permitting, matter can nonetheless conspire into living cellular conglomerations that manage, at least locally, to reduce or assuage entropy. How is that great trick pulled off? How does life manage to overcome the unfeeling specter of mathematically determined probability? How is the incessant tide of entropy co-opted into doing useful work? To answer this we need to think a little more deeply about living things.

As we know from experience, any living organism maintains order, managing to perpetuate a state of being far from equilibrium. This is rather like an enduring vortex within a stream of water, or even like the giant red spot of Jupiter, which has been observed by astronomers for hundreds of years and which is in fact a giant and persistent vortex of gas—another example of an island of order, or robust pattern, persisting in a sea of entropic disorder.

The ability of living things to maintain their structural integrity in the seemingly harsh face of entropy is referred to by biologists as *autopoiesis,* a word that literally means "self-making" or "self-generating." Autopoiesis is a hallmark of life and its continual evolution (if a sugar cube aspired to autopoiesis, it would have to frantically rebuild itself after being tossed into a mug of hot coffee). Our bodies, for example, are rife with autopoietic self-generation. As Fritjof Capra notes:

> Our pancreas replaces most of its cells every twenty-four hours, our stomach lining every three days; our white blood cells are renewed in ten days and ninety-eight percent of the protein in our brain is turned over in less than one month. Even more amazingly, our skin replaces its cells at the rate of 100,000 cells per minute.[2]

A visual image that comes to mind in thinking about autopoiesis is that of a serpent eternally eating its own tail. Such a system feeds off itself, driving itself into a self-perpetuating state of being. In this sense, the evolution of life represents the continuous creation and sustenance of autopoietic systems, systems that are about as far away from entropic equilibrium as we can imagine. As biologist Lynn Margulis and her son Dorion Sagan explain in their wonderful book *What Is Life?*:

An autopoietic entity metabolizes continuously; it perpetuates itself through chemical activity, the movement of molecules. Autopoiesis entails energy expenditure and the making of messes [excretion]. Autopoiesis, indeed, is detectable by that incessant life chemistry and energy flow which is metabolism. Only cells, organisms made of cells, and biospheres made of organisms are autopoietic and can metabolize.[3]

But there is a price to be paid. While organisms can stave off the current of entropy (at least until they die), they do so by exporting entropy or disorder into their environment. This means that although an organism appears to maintain order in the face of entropy, if we take into account the organism's immediate environment, then entropy is still increasing. The increase in entropy is in the form of waste products—both solid wastes and heat. The same principle of a continual flow of matter is also found in fluid vortices, which import fresh water molecules and export others while maintaining their overall pattern. You can observe this yourself the next time you watch the bathwater go down the drain. In a real sense, the enduring and self-replenishing water vortex exhibits a basic kind of autopoiesis, though obviously not as complex as the autopoiesis found in living organisms.

More pertinent examples come to mind. For instance, we eat an apple, a highly organized configuration of matter, and later excrete what our bodies do not need. We have thus extracted some useful energy/order from the apple but have, in the process, produced a small amount of entropy. All organisms excrete in some way. Material of some kind is absorbed—usually with a high degree of order, pattern, and potential energy—and less ordered and less energized material is excreted. Again, we can see that an organism is akin in principle to a stable vortical pattern within running water, absorbing organized material in one end and exporting disorganized material at the other. By making use of the order or high-grade energy in food and by excreting waste, organisms thereby manage to remain intact and preserve their unique structural pattern.

Organisms do not merely excrete waste in the forms with which we are scatologically familiar; they also excrete heat. Lots of it. Heat is an inevitable waste product of metabolism. Being molecular motion, heat is a form of energy that, in general, cannot be reused. Once energy has been converted into heat, statistical law ensures that such molecular vibration cannot be reclaimed to do useful work. Heat invariably spreads out and dissipates. With mammals, infrared cameras reveal the aura of heat that they emit: a constant envelope of heat emanating as entropic waste from metabolic processes. It is the same with systems like machines. Touch the TV set after it has been on for a few hours and one can discern the heat being exported into the environment. This also explains why computers have heat sinks and fans inside them—they serve to dissipate the heat produced when the computer is being used. The same is true of all machines. No machine is one hundred percent efficient in its use of energy, because some energy is always lost as heat. Quite literally, heat production is a waste of energy in both organism and machine.

Even while we eat and digest an apple, the combined metabolic work of our cells will result in more heat being exported into our surroundings than is the case when we are not eating and digesting. This radiated heat, or the slight increase in the molecular vibration of air molecules, is probably *the* classic form of entropy. We cannot harness radiated heat and turn it back into some useful form of energy. In effect, this means that we cannot recover the molecular motion that is heat energy and convert it back into a material form of energy. So, although energy is always conserved in Nature, this energy can be in different forms. Some primary high-grade forms—such as light from the sun—can be sequestered into doing interesting constructive work (as evidenced in photosynthesis in plants, wherein sunlight is converted into food), whereas in the form of heat it merely adds to the inexorable increase of entropy. This is the price of life and its evolution. While the biosphere evolves and builds up ever more refined systems of autopoietic order (organisms and ecosystems), entropy, in the form of heat, is exported into space.

Having said all that, an inevitable increase in entropy is not a bad price to be paid for the existence of life and its evolution. Indeed, Nature has struck a kind of bargain with itself. The linkage of entropy with life is, as Earth scientist James Lovelock has pointed out, akin to the necessity of using up flashlight batteries to see in the dark. Without the use of energy, nothing can be seen or done. If we wish to use the flashlight, we must sacrifice some energy (increase entropy) in the process. But what is really remarkable is *the very existence of exquisitely ordered living things.* Evolution works. Brilliantly so. Going against the grain of entropy (at least locally), pushing against the general statistical tendency of the Universe to run down and reach a deathly state of equilibrium, matter nonetheless conspires to organize itself, thereby producing the speciated wonders of the living world.

ENTROPY

A Life-and-Death Situation

The tendency for things to run down ensures that organisms eventually die. Actually, there is one theory that holds that the reason we age and die is because of the buildup of random errors that occur during cell replication. Because our cells constantly replace themselves, there is inevitably some replicational inaccuracy and it is possible that this buildup of copy errors manifests itself as aging and the movement away from perfect physical health. Once again, this is an entropic process, rather like making repeated photocopies of an original print, each time using the copy just made. Errors creep in and the original pristine image gradually becomes less defined. Old flesh is corrupted flesh. Of course, by reproducing beforehand, life carries on in spite of death. Obviously, the promulgation of sex cells is accurate enough to allow healthy offspring to be created before entropy swings its fatal sickle. Therefore species survive, while individuals do not. And even if a species dies, life as a whole continues.

Likewise our machines run down. They wear out. One by one the constituent components of, say, a fridge will, in time, decompose and the fridge will be no more. This applies to any useful system we build, be it a TV, a car, or a computer. Like organisms, machines are patterned systems constructed by the careful application of energy, and each runs by carefully using some form of energy. Yet all cars, all computers, and all televisions will not remain unsullied and functional forever. All machines wear out in the long run. No object, man-made or otherwise, is static. Everything is subject to the entropic ravages of time. Take any artifact, say a car, and place it in a desert. Over the years the car will decay through constant weathering. Allow a million years to pass and all that will remain will be a few scraps of plastic and rusty (i.e., oxidized) metal. Organized systems such as machines cannot survive indefinitely. Artifacts do not repair themselves, nor do they reproduce.

Contrast this with the various organisms that make up the tree of life. There is a subtle and brilliant difference. In the imaginary desert in which we place our car and leave it to decay over the millennia, we may find, within the sand, a small ant colony. Whereas the car will disintegrate over the years, our ant colony should surely remain. Of course, the individual ants themselves will not be the same, yet the *species itself* manages to survive. The life of species and patterns of successful behavior do appear to survive the entropic deconstructional effects of time. This is a seemingly simple observation, of course, but it nonetheless highlights the fact that evolution is a very *clever* process (i.e., smart and dexterous). It is also the case that, given enough time, the ordered pattern embodied by the ant species might well transcend itself by evolving into two subspecies or even a new species. So, far from decaying, the ants and their autopoietic capacity actually manage to evolve and further complexify—potentially, at least.

So not only has evolution managed to build immensely complex structures, but when those individual structures (organisms) finally succumb to the more usual flow of entropy, the greater system of life nonetheless prevails and wins the day. The autopoietic thrust of life is assuredly dogged and tenacious, so much so that the biosphere has withstood fully three and a half billion years of time, each second of which constitutes the ever-present "threat" of dissolution. It seems then that not only can evolution create self-sustaining ordered patterns, *it also manages to get better and better at doing so.* As life continues feeding off itself and continues to stimulate its own autopoietic nature, so too does it become more and more ordered, more and more complex, more and more patterned.

This, then, is the quintessence of life, the crux of the evolutionary process we are born into and which we take so much for granted. Life (and its evolution) is a self-maintaining, self-generating autopoietic process constantly staving off disorder and, against all the odds, somehow managing to perpetuate itself in quite spectacular forms such as that of the hummingbird, that of the jaguar, or even that of humans. Cells die, organisms die, even species eventually die, but *life itself* continues to be. Against the decadent tides of disorder, Nature weaves together basic units of matter into macroscopic conglomerations that, in terms of their underlying pattern and behavior, are remarkably far from equilibrium. Once more think of an unsullied human and a liquidized human. Evolution is the force that makes the vast structural difference between dead slops of matter and integrated living organisms. In essence, life is an exquisite and enduring organismic pattern rising up and out of a Universe that is running down (like a watermill using an entropic downward water current to power itself upward). This spectacular unfolding potential is a clear indication that Nature is endowed with intelligence, purpose, and meaning.

THE MYSTERY OF BEING

In a way, it is as if the Universe poses a problem to itself vis-à-vis entropy. By statistical law, the entire gamut of reality is inevitably running down, wasting away, apparently doomed to flicker out in a pitiful heat death, all its energy dissipated into featureless molecular vibration. By the constant random mechanical shuffling of its parts, the Universe is heading more and more toward thermal equilibrium. How to stop this inherent running down, this slow but relentless death? The solution, it appears, is embodied in the ongoing evolution of the tree of life (wherever one should arise). That Nature can pull off a crafty thermodynamic trick or two is exemplified in every one of the ten billion bacteria in a pinch of soil, in the billions of insects scuttling throughout the biosphere, and in the 700,000 or so plant species that now exist. Organisms work, they work well, and, en masse, they feed back on themselves so as to stimulate the evolution of ever more refined and complex species, and even consciousness. But more of this later; for now, it is enough to acknowledge that life and its evolution be a localized move away from the general tendency of the Universe to simply run down, or at least that life is a constructive response to the inexorable entropic "decline" of our sun. Moreover, this autopoietic imperative of Nature, this remarkably creative use of our sun's output, has led to you and me, for here we are.

The question we simply have to consider once more is "why?" Why does Nature *perform* in this kind of way? Where did the *capacity for evolution* actually come from? Indeed, while we may acknowledge that evolution can explain the origin of species, how do we explain *the origin of evolution itself?* The same applies to the genetic code—why is it possible and why does it work so well? In other words, what are the chief factors or chief conditions that must hold sway such that the evolution of complex patterns can become manifest? For if entropy is mathematically determined, where did the obviously good odds of life come from? (Life is now seen as a "cosmic imperative," to use the words of biochemistry professor and Nobel laureate Christian de Duve, whom we shall hear more from in a later chapter.) It is here, in the answers commonly provided by orthodox science, that we can again sense a kind of conceptual myopia. For the paradigm commonly taught with regard to the sense and significance of this ultra-magnificent process we call evolution holds that there *is no reason or purpose whatsoever for the evolution of life,* only that it just happens, and that there is nothing much more to say about it.

If we truly believe that the evolution of the tree of life is but a pointless brute factual capacity of Nature, then we truly are condemned to live and die in a kind of prison, a senseless and accidental prison upon whose existentialist walls we may inscribe some cultural meaning or another, but a prison nonetheless devoid of any real meaning apart from the meanings we invent. True, we might marvel over the wonder of life, gasp in amazement at the sculpting power of evolution, yet, at heart, it is but a pointless incident. After three and a half billion years of frenetic autopoietic evolution, consciousness is thrown out onto the stage and wonders at the beauty of it all, only to be told by science that evolution "just is," that evolution has no direction, no intelligence, no meaningful agenda, no nothing. It merely happens, lumbering on like an aimless freewheeling snowball. Autopoietic patterns similarly "just happen." And although all this "happens" rather magnificently and with spectacular results, any meaning or purpose we might divine in the operations of evolution is but wishful thinking. To quote iconic evolutionary biologist Richard Dawkins: "The universe we observe has precisely the properties we should expect if there is, at bottom, no design, no purpose, no evil and no good, nothing but blind, pitiless indifference."[4]

IS REPRODUCTION THE NAME OF THE GAME?

If the evolution of the autopoietic tree of life does indeed represent little more than a meaningless affair with no purpose, reason, or inherent intelligence, then the sole reason we exist is not to marvel over the Universe but to reproduce. Having a good time, reading, musing, philosophizing, aspiring to poetry, embarking upon some career or spiritual quest—all this is beside the point. Reproduction is the sole priority and any other interests are secondary to this. To biologists like Dawkins (who popularized the notion of blind selfish genes), organisms are little more than elaborate vehicles through which genes replicate themselves. Driven by genetic code, autopoietic systems like organisms might be all well and good, representative of an admirable creative capacity of evolution to rise against the tide of entropy, yet it is quite pointless really. Somehow, life started. Once unicellular systems were up and running, evolution dictated that the better any of these systems was in its ability to reproduce itself, the more it prospered, the more it ousted more inferior systems. Reiterate this logical process and you have life as we know it: splendidly mindless, splendidly devoid of intelligence.

According to this prevailing paradigm, humans, despite their extensive brains and their conscious intelligence, represent just another organismic technique for certain genes to replicate themselves successfully. In this sense, one's gonads and the ability to make fertile use of one's gonads are indeed the only thing we are good for. The joke about men being the way sperm get to replicate themselves and women the way ova get to replicate themselves is, in the above paradigm, a tad close to the mark. We can talk, muse, and build cultures, but these are mere diversions, sort of sideline activities or props that furnish the real motive of existence. And the real motive of life is to fulfill the evolutionary imperative of going forth and multiplying. That is our only true obligation, for it ensures the survival of our genes and thus our existential compliance to evolutionary logic. If our ancestors had neglected to carry out this primary evolutionary dictate, then they would not have been our ancestors. Thus, reproduction is all we are really good for.

While being compelling in terms of cold hard logic and orthodox thinking, this selfish gene paradigm is pretty bleak. No one can deny it. Are we nothing but walking, talking, gesticulating, flirting gonads? Are we really no more than colorful *excursions* in the art of genetic manipulation and genetic replication? Do our trillion-celled bodies exist merely for the purpose of safeguarding and copying their genetic quota? Do we live and die solely to satisfy the strict evolutionary logic of replication?

That is what science currently teaches us. Of course, one might argue that although we are here merely to pass on our genes, we can enjoy culture and art and family and friends and such along the way. We can build huge monuments, construct elaborate philosophies, compose symphonies, write books, make poignant music, dance, formulate ethical beliefs, cook delicious food, enjoy all manner of things, and even enjoy sex without incurring offspring until we wish to do so. But in the context of a paradigm that holds that evolution is pointless and that Nature is a non-intelligent brute factual system that just happens to be blessed with incredible powers of biological articulation, all these human activities become but temporal nonsense. In a big enough context, all this human grandeur, from the pyramids to Bach to the conquest of space, fade into obscurity, just a lively yet essentially meaningless and anomalous blip in the general entropic decline of the Universe. If we are mere pawns in a cellular game of replication, all that matters is getting our gonads fused to a member of the opposite sex, thereby ensuring the survival of our genes. We are in the game of life for fourscore years and ten and then dispatched, replaced by offspring pawns who inherit our genes and repeat the process. And so on and so on ad infinitum . . .

It is a wee bit grave, all this. A dire paradigm really, for life emerges as the most outlandishly pointless phenomenon we can imagine. It's all bells and whistles with nothing of real significance underneath. Even if we found out that we were the product of some divine ruse, at least that would be something, for it would indicate that there was a certain contriving intelligence underlying our existence. But this is not the case, so we are told. The only meaning in the Universe is the meaning *we* invent. The only high-level intelligence is *our own,* with maybe ET, dolphins, and our pets for company. There is no greater intelligence than ours and that of the artificially intelligent machines we invent. There is nothing greater than us, no bigger meaning to be grasped in the Universe at large. At deep dark bottom, Nature is utterly devoid of purpose, evolution an explosive incident of happenstance.

This, then, is the dominant paradigm that we are up against. Even if it is not taught quite so dramatically, a meaningless and mindless view of Nature is assuredly implicit within the rhetoric of mainstream evolutionary science. However, this paradigm is certainly (and thankfully) on its last legs and is not destined to persist much longer. The alternative paradigm revolves around the concept of natural intelligence, according to which life is a self-organizing intelligence, an ultra-smart technology if you will, whose source is rooted in the very laws of Nature. As we shall see, there is much evidence to support this new and innervating take on life.

CHAPTER 3

∽

EVOLUTION

Never Mind the Gonads, Here's the Real Agenda

It is time to look even more closely and critically at the orthodox evolutionary paradigm we are attempting to overwrite. To recapitulate, it is not that this paradigm is wrong, only that it is not right enough. It is not the whole truth. It is not the whole story. Obsessed with details like individual genes, the orthodox paradigm fails to capture the confluential panache and sheer organismic acumen of life, serving instead to reduce organisms and the sophisticated feats of biological engineering that they embody to merely this or merely that genetic mechanism. This is reductionism gone awry, the zealous urge to reduce autopoietic artistry into bits and pieces seemingly explicable in their own terms without reference to any bigger context. While we would never dream of reducing an acclaimed painting or a musical symphony in the same reductionistic way, Nature's tree of life is treated differently.

The creed of merelyism tells us a lot about human psychology. It is clear that the majority of scientists and philosophers like to consider themselves and other humans as the sole possessors of "high" intelligence, whereas the process that built them (i.e., evolution or Nature) is considered pointless and mindless. Therefore, any paradigm that bolsters this kind of merelyistic egoism is destined for a warm reception. The more the complexity of life is reduced, the more smugly confident some may feel about themselves. The more we can remove signs of intelligence operating within the evolving tree of life, the more unique and esteemed human intelligence becomes. It's no surprise then that merelyism attracts research grants and academic acclaim. Unfortunately, the same cannot be said for alternative and more holistic approaches to understanding life. As we shall see in this chapter, a more holistic approach means an approach in which context (i.e., the sensible environment and the sensible laws of Nature operating therein) is the key factor with which to understand both the origin and subsequent direction of evolution.

In order to allow the new paradigm of natural intelligence to acquire momentum, let us begin by taking a basic example of a living organism and seeing how the old paradigm is applied so as to explain that organism's existence. According to the thought experiment of the last chapter, we have already established that organisms represent significantly arranged systems/patterns of atoms, molecules, and cells. Now we need to see in even more detail how merelyism explains such evolutionary feats of organic organization. As alluded, genes are the key focus. A tiger will serve as our study example. Consider one in your mind's eye.

Now, any sane, sober, and right-minded person gazing upon a tiger will doubtless be impressed by the apparent design, for a tiger is an exceptionally beautiful creature capable of running, hunting, calculating, digesting food, excreting waste and toxins, breathing, maintaining its internal temperature, healing itself, cleaning itself, reproducing, and so on. Flaunting beautiful autopoiesis, a tiger has what it takes to survive. Well, apart from the relentless threat of man, that is. Notwithstanding the threat of our species, tigers are patently highly efficient living organisms.

As we have ascertained, organisms like tigers are constructed from trillions of cells: skin cells, liver cells, brain cells, muscle cells, and so forth. These cells are themselves made up of trillions of molecules, and those molecules are made from atoms. A tiger is thus a singularly significant arrangement of physical material, untold trillions of basic parts orchestrated into a coherent enduring pattern. In short, a tiger has striking design (of some kind) written in it and all over it. It's no wonder, then, that people in the past were tempted to see organisms as the work of a divine hand. But we live in more educated times. From our vantage point atop centuries of science, we are able to build upon Darwin's revolutionary insights and see that multicellular organisms like tigers are not wrought ex nihilo by the hand of God almighty but are the manifest result of evolution through natural selection (i.e., the hand of "almighty" Nature). Natural design, in other words. We have at least to buy into this idea before proceeding with our analysis.

GENES AND DNA

Biological science tells us that the physical form of a tiger and its behavioral repertoire (its phenotype) are determined by its genes, those long sequences of DNA residing in each of its cells. We won't take issue with this either. All the evidence points to the fact that DNA is a digital code (quaternary and not binary like computer code) bearing instructions on how to link together amino acids so as to make proteins, which, in turn, make cells, tissues, and organs. Genes, being discrete segments of DNA that get replicated during reproduction, are thus sets of instructions for eliciting biological autopoiesis—they code for the components that, when organized, elicit the sort of nifty biological processes previously considered. In the tiger, there may be gene complexes that control musculature or the immune system or fur coloration and so on.

The plethora of genes that exist within the tiger gene pool are precisely those genes that work, those that confer upon tigers morphological and behavioral characteristics that serve to enhance the reproductive success of tigers. Genes for building strong leg muscles will obviously be passed on to future generations since tigers with strong leg muscles will be able to run faster and thus hunt better. More food equals better fitness equals better reproductive viability. A gene that made a tiger lopsided or extremely timid in the presence of herbivores would not survive long, because such a gene would obviously be selected against. So it is that the hand of Nature continually selects among various genes (i.e., heritable and potentially variable strands of DNA code). Those genes bearing instructions that aid the genes' own success in replication are favored by Nature, whereas less survival-favoring genes are selected against. This is all armchair stuff. Indeed, one can reason in this way about natural selection without actually having seen a tiger. Moreover, the reasoning can be applied to any organism you care to imagine.

So where did genes come from? If tiger genes are nifty segments of digital information that serve to facilitate the creation of highly organized tissues, organs, and behaviors, then where did this rather striking digital code originate? In other words, how did a tiger and its unique quota of genetic material arise? The short answer is that genes and groups of genes change and evolve over time. A DNA-writ tiger did not just spring into existence fully formed. Somewhere along the long line of historical time the tiger evolved, or descended, from a creature similar to itself but not quite the same (different enough to warrant calling it a different species). And this ancestral species itself descended from a still earlier species. And so on all the way back to the effervescent primeval soup. Recall that Darwin's prescient book was titled *The Origin of Species,* indicative of its detailed explanation of how new species can emerge over time. Once you have a genetic system (i.e., heritable information), then, through variation and natural selection, the information content of the genetic system will change over time, and thus new species will invariably stem from one another: hence the evolving tree of life with its myriad branches, twigs, stems, and shoots; hence also the well-worn dogma that evolution is nothing more than a change in gene frequencies over time.

At first blush, this account of the origin of the tree of life might still seem hard to swallow, yet it is undoubtedly true—at least up to a point.

As long as we bear in mind that the tree of life has been growing for more than three and a half billion years, then we can conceive that small, almost negligible changes have always been occurring within the genes of creatures. The gene pool of a species—its collective genes—is never static. There is always variation of one sort or another, caused either by mutagenic forces, sexual mixing, or random errors in gene replication. Variation in genetic material is critical for evolution, for without variation there can be no selective change, and it is precisely non-random selective change that eventually yields entirely new species. Given enough time, not only will the gene pool of a species vary, but eventually variant groups will emerge different enough from one another such that the two groups are no longer capable of mating. This implies the origin of new species and new genera.

Selective variation is a cumulative step-by-step process; it's very, very time-consuming, but since time is a vast cosmic commodity, it matters not how long selective variation takes. This slowness explains why we cannot generally observe the formation of actual new species. However, we can observe basic selective variation in action. Think of dog breeding, for example, where all types of dogs are variants of a single ancestral species of wolf. The myriad types of dogs within our culture have all resulted from humans selectively breeding certain characteristics from the canine gene pool. Indeed, one could even argue that since certain dog breeds are so different in size, structure, and temperament as to negate effective mating, then, for all intents and purposes, such breeds are actually new species, or at least they represent, in Darwin's words, incipient species. Of course, whether or not we do decide that a particular breed represents a new species is up to the discretion of taxonomists (and they are forever arguing about these issues), but the point remains that genes have the potential for constructive variation and are selectively heritable.

If we apply this mechanism of slowly evolving genetic change to our own species, we have to admit that modern humans—*Homo sapiens*—evolved from ancestral hominid species (like *Homo erectus*). Go back in time far enough and we are related to some common ancestor of all currently existing primates. Indeed, this adequately explains why fully 98 percent of a chimpanzee's DNA is the same as ours. And primates themselves descended from still earlier mammalian creatures, which explains why we share genes with creatures as different from us as nematode worms. Unsurprisingly, people found this evolutionary logic hard to swallow in naive and puritanical Victorian times, yet this is the implicit message of Darwin's groundbreaking insights. For all man's present attributes, we are but one particular primate species descended from a long line of ancestral primate species now extinct. Apes, great or otherwise, are our distant cousins. It is quite a humbling notion, highlighting our intimate connection to the rest of life's great web.

THE APPARENTLY MINDLESS ART OF REPLICATION

Despite its "sluggish" progress and essential invisibility, the principle of natural selection wherein new species emerge can be simply grasped. It is a logical consequence of non-random DNA replication. Once you have a genetic system utilizing DNA, then genes that work, that is, genes that serve to promote their own replication by building life-affirming body structures and life-affirming behaviors, will prosper and flourish. Organisms like tigers can therefore be viewed as strikingly elaborate vehicles for gene replication. Indeed, species *are* considered in this way by evolutionary biologists. This, in fact, is the bottom line of the dogma. As Richard Dawkins boldly asserts: "The great universal Utility Function, the quantity that is being diligently maximized in every cranny of the living world is, in every case, the survival of . . . DNA."[1]

Dawkins's use of the term "Utility Function" is a way of pinpointing the most essential principle (for him) operating throughout the living world. As stated, this most essential principle is ostensibly the replication of genes, of heritable strands of DNA, that code for useful characteristics like muscles, strong wings, sexual appeal, alertness, and so forth.

The original bits of replicating DNA (or RNA) three and a half billion years ago might have been relatively simple—they just copied themselves without constructing any elaborate body parts. But those coils of DNA that could utilize resources more efficiently by building body structures—via random alterations in their DNA caused by mutation and other forms of variation—would, by definition, have a distinct reproductive advantage and thus gradually outnumber other variants. Reiterate this generate/test/generate/test process billions of times and you can see that DNA becomes (or can become) more and more complex as it clothes itself with more and more elaborate body parts for its own replicational ends. Over large spans of time, the biological patterns induced within the evolving tree of life become richer and more pronounced.

That, in yet another nutshell, is evolution, as we know it. The compelling logic of "selfish genes" is as hard and as cold as tempered steel. The evolution of life is, apparently, nothing more than the mindless replication of bits of DNA filtered, sifted, and sorted by Nature, the hand of which selects precisely those sequences of DNA that act to replicate themselves successfully. The result is species, each of which is adept at promulgating genetic instructions.

Back to the tiger . . . or what's left of it. What we have now is a superbly successful replicator, essentially an organic utilitarian vehicle of flesh and bone transporting genes for successfully making copies of itself. All organisms can be treated in the same blunt and impassionate way. Every species of life on Earth, whether insectile, botanical, or animal, is a particular method whereby certain groups of genes manage to replicate themselves. No matter what kingdom organisms belong to, no matter what sort of basic strategy they use to survive and reproduce, all utilize instructions written in digital DNA in order to regenerate and reproduce. Gene machines one and all. Any poetic inclinations we may have toward a tiger are thus beside the point. If a tiger's eyes burn brightly in the night, that is only because certain tiger genes code for moist spherical eyes. A tiger is "merely" a gene machine descended from billions of ancestral gene machines all the way back to the first simple strand of replicating DNA (or RNA). And that's all there is to it. Good night and God bless.

Not quite, actually. But this is where we have to get a bit more discerning in our analysis. We have to consider the above process of genetic replication with a little more subtlety. Let's start by considering in more detail Dawkins's obsession with isolated genetic details.

SQUARING UP TO THE
BLIND WATCHMAKER'S APPRENTICE

Richard Dawkins is undoubtedly one of the world's foremost defenders of Darwinian orthodoxy. He is, as they might say on the streets of East London, the "Daddy" of modern evolutionary thought. Exceptionally proficient in the art of popularizing scientific ideas, Dawkins has influenced an entire generation of people interested in biological evolution. His best-selling books can even be found in supermarkets alongside tawdry celebrity tomes. However, many people dislike Dawkins—none more so than the theological types who see themselves as ambassadors of the Lord and whose religious sensibilities are assuredly undermined by Dawkins's alarmingly candid rhetoric. Such people nip at his heels and stab limp rubbery knives into his ankles. They brandish crucifixes and hurl lines of scripture at him as if he were Dracula incarnate out to suck them dry of their precious religious faith.

Appealing to their own sense of incredulity at evolutionary logic, critics—especially theological critics—have not a hope in hell of dissuading Dawkins and his supporters from the "selfish gene" account of life. Dawkins's eloquent articulation, the sheer logic of his argumentation, wipes away such adversaries. Indeed, in the domain of popular science, Dawkins has now become a kind of world power. Armed with his particular brand of merelyistic logic, which has no place for natural intelligence, Commander Dawkins now heads a vast army of atheistic devotees, his wake strewn with the writhing bodies of felled theologists and creationists unable to come to terms with his notion that complex life is the outcome of little more than a godless "blind watchmaker."

SYSTEMS THAT MAKE SENSE

At one point in his influential book *Climbing Mount Improbable* (Dawkins has written numerous acclaimed books about evolution, all with the same explicit message), Dawkins garners support for his defense of Darwinian orthodoxy by focusing upon a particularly striking experiment. We shall consider this prototypical experiment in depth since it highlights in a most obliging way a certain fundamental principle at work in evolution not acknowledged by either Dawkins or the original experimenters. Not only is this unattested principle involved in the evolutionary processes we have been discussing, it also leads us into the sublime realm of natural intelligence.

The principle in question is connected with *context,* an incredibly important factor that is the very key to the new paradigm. For to invoke context in discussions of natural selection is to give equal importance to the contextual environment in which genetic information is changing. Once contexts are invoked, it becomes clear that Nature operates as a contextual feedback system, moreover a specifically configured feedback system able to provoke not simply a change in genetic information over time but *sensible changes* in genetic information over time. Because Nature represents a sensible and intelligible context, this means it can be made sense of and can be learned about. Thus, through trial and error, life can learn more and more about how to make sense of the sensible context in which it is embedded and how to utilize that learning so as to sustain itself. Genes, of course, are conveyors of that learned information. Furthermore, if a system like Nature is configured and ordered in a sensible and intelligible way, then, as far as I can tell, intelligence of some kind must be bound up with it (inasmuch as intelligence and intelligibility are bound together). If so, the evolution of life can be seen as a translation of this ambient intelligence into a new form. Trust me, this will become much clearer as we continue.

Back to the chase. The experiment under scrutiny here concerned the evolution of a virtual eye within the simulated reality space of a computer system. This kind of experimentation is not so unusual at the current time since, as we shall discover in the next chapter, more and more evolutionary research is utilizing computer-based virtual environments. Researchers explore things called genetic algorithms, which are computer programs that allow evolutionary processes to unfold. The aim is basically to "cook up" the principal procedural ingredients of evolution within the computing space of a computer system. This is achieved by writing clever software that creates a virtual environment with its own laws that, like natural laws, serve to promote evolutionary processes. Genetic algorithm is a generic name for this kind of potentially creative software program.

Utilizing genetic algorithms, digital entities (i.e., entities underscored with digital computer code as opposed to digital DNA) can evolve within a purely virtual environment as opposed to a more tangible environment. The fate of these virtual entities can be determined according, say, to how able they are to gather whatever "resources" are available in the virtual environment they find themselves in. This represents a measurement of their virtual "fitness." The more fit the digital entity, the better able it is to reproduce and thus spread the computer code of which it is constructed, which codes for its advantageous behavior. In this way a process analogous to natural selection can transpire.

The really good thing about computers is that they operate extremely fast, allowing millions of generations of virtual organisms to be bred in the space of a few hours. Unlike real life, you don't have to hang around for "Cambrian ages of time" in order to witness significant evolutionary results. Poor old Mendel, the grandfather of genetics, had it hard in his day, pottering about his monasterial garden, planting peas and divining the slow workings of evolution and heredity. No super-fast computer for him with which to quick-breed virtual peas during his lunch hour.

Anyhow, in this particular virtual experiment, the scientists involved wanted to know whether something as doggedly complex as an eye could be evolved from very humble beginnings. The eye has always vexed those who want to grasp evolution in its most fantastic manipulations. It even vexed Darwin, who saw in the sophistication of the eye a challenge to his theory of gradual step-by-step evolution. In *The Origin of Species,* Darwin bemoans:

> To suppose that the eye, with all its inimitable contrivances for adjusting the focus to different distances, for admitting different amounts of light, and for the correction of spherical and chromatic aberration, could have been formed by natural selection, seems, I freely confess, absurd in the highest degree.[2]

Darwin is being a bit melodramatic here, for he must have well realized that the eye, in terms of complexity and sophistication, is really no more remarkable a feat of natural design than is the liver or the inner ear or the human cortex (one wonders what Darwin would have made of the frenetic biochemical processes going on inside a cell had he been privy to them—these processes are astounding in terms of complexity and purpose). All these organs, any organ in fact, should command our respect and admiration. But since the eye is the chief means through which we sense the world, it is no wonder that we marvel more over its design than that of other organs. If only Darwin had had access to a time machine with which to visit the computing labs of the scientists we are discussing, for then he would have appreciated their experiment and its outcome. For nothing less than the simulated evolution of an eye was accomplished.

The scientists commenced by placing within the computing space of the computer a population of very simple "virtual eyes," consisting of flat layers of light-sensitive cells sandwiched between two other equally flat basic layers. Underlying the structure of these virtual eyes was computer code, this being somewhat analogous to genetic code, especially since the code could be altered. Needless to say, the "light" within the virtual environment was not real light but a form of virtual light, virtual rays of digital information that were lightlike, adhering to lightlike principles of manifestation to which a simple layer of virtual light-sensitive cells would react. Thus, these original simple virtual eyes were able to focus some of the virtual light, their success in doing so being measurable.

To reiterate in Dawkins's words, the scientists' starting point was little more than

> a flat layer of photocells, sitting on a flat backing screen and topped by a flat layer of transparent tissue. [The scientists] assumed that mutation works by causing a small percentage change in the size of something, for example a small percentage decrease in the thickness of the transparent layer, or a small percentage increase in the refractive index of a local region of the transparent layer.[3]

What the researchers did was insert a population of these virtual eyes into the program and then start randomly altering their underlying digital code in very small ways. This was done in order to mimic the action of mutation and random variation in the real world, the actual amount of variation being based upon real-world mutation rate data. This process led to minuscule changes in things like the refractive index of the transparent layer, as noted by Dawkins (a refractive index concerns the degree to which light rays are bent). The resulting second generation of virtual eyes was then measured in terms of their accuracy in focusing virtual light rays. Those virtual eyes that through chance variation were better able to focus simulated light, to bend light rays into concentrated focus, were granted a very slight reproductive advantage in terms of reproductive success. A new generation was then bred from this second-generation stock, taking into account the slight reproductive advantage held by a minority of them. The test-and-select procedure was again applied. And so on thousands and thousands of times. Until . . .

A veritable miracle occurred! Eventually it was found that a "fish eye" evolved, having formed itself out of the transparent layer. From a poor slipshod beginning of just a flat three-layered system, a round fish eye—complete with lens—had evolved, which was able to focus light in quite a refined manner. The lens had actually evolved by gradual point-by-point changes in its transparent layer. And it all happened on its own! That crown of biological creation, an eye (at least a simulated one), had formed itself within a computer-generated virtual reality, thus demonstrating that ordered and complex structures can be generated "merely" by the reiterative action of random mutation and non-random differential reproduction. The experiment confirmed what Darwin himself actually suspected about the evolution of the eye despite his initial incredulity that natural selection alone could account for it. A little after the preceding quote, he wrote:

> If we must compare the eye to an optical instrument, we ought in imagination to take a thick layer of transparent tissue, with a nerve sensitive to light beneath, and then suppose every part of this layer to be continually changing slowly in density, so as to separate into layers of different densities and thicknesses, placed at different distances from each other, and with the surfaces of each layer slowly changing in form. Further we must suppose there is a power always intently watching each slight accidental alteration in the transparent layers; and carefully selecting each alteration, which, under varied circumstances, may in any way, or in any degree, tend to produce a more distinct image. We must suppose each new state of the instrument to be multiplied by the million; and each to be preserved till a better be produced, and then the old ones to be destroyed.[4]

Darwin's prescience is truly uncanny. One suspects that he *did* have a time machine, and that he secretly visited the science lab where this simulation took place. Unable to understand all the computer jargon, he then termed computer technology a "power" and duly translated his observations into the eloquent but non-computational language of his time. Or maybe he had a precognitive dream about the experiment. At any rate, like Dawkins, we should be most impressed with this computerized simulation of evolution. If an apparently "dumb" computer program can generate a virtual eye complete with lens, then surely the action of evolution by mutation and variation in the real world could similarly produce real eyes. And tigers, of course. The whole gamut of life, in fact.

What this classic experiment purports to highlight is that extraordinarily complex things like eyes—and, by extension, the very tree of life itself—can indeed gradually evolve over time as long as there are mutational changes occurring within the coding system that codes for organic form, and that these discrete changes in the coding system, even though they are really small, can nonetheless accrue and eventually amount to incredibly creative effects. Actually, this must be true, since once you have a population of code-driven entities whose behavior is subject to some kind of measurable success, a small amount of variants will undoubtedly perform better and these will therefore become selected and survive. In the real world, of course, the code subject to cumulative heritable change is genetic DNA, while in the simulated evolution just discussed, the code subject to cumulative selection was the digital binary code that coded for the virtual eye entities. In each case, the result is the same. Evolution pure and simple. *Quod erat demonstrandum.* The computer simulation can be turned off, and we can leave the lab confident in the knowledge that there is really nothing to evolution at all. It is simply a matter of time, chance, and circumstance, no more than a change in some kind of information/gene pool over time. From very little we can get major significant things. Something from nothing. Apparently.

What Dawkins fails to note about this important experiment, however, is the fact that one is *not* deriving meaningful and impressively designed structures out of thin air. This is what *seems* to be happening, for without the intervention of humans an autonomous computer program still manages to produce (i.e., design) a striking object, namely a fish eye endowed with a neat lens. It would appear that the scientists simply set up the program, fed in the initially flat and lensless virtual eyes, and then left for their lunch. On returning they found, much to their satisfaction, that a round fish eye had evolved. It just happened of its own accord. Indeed, it would seem that blind dumb luck was the only force at work. Thus we seem to be getting meaningful patterns and significant structures arising from blind and essentially meaningless processes. However, this conclusion is erroneous. Why? Because in each situation, whether it be a fish eye evolving in virtual reality or a tiger evolving in the real world, *it is the presence of a sensibly ordered context to start with that allows sensibly ordered things to arise.* This is an indisputable point intimately bound up with evolution, although it is rarely discussed. Yet it is of prime importance if we wish to understand the way in which the tree of life has been carefully watered and fed.

CONTEXT

The Missing Link to Natural Intelligence

An a priori sensibly ordered context is assuredly one of the most crucial factors involved in any evolutionary process. In order for the virtual eye experiment to have worked in such dramatic fashion, we find that a sensibly ordered context was needed so as to *provoke* and literally *draw out* the round fish eye. This sensibly ordered context was the particular virtual environment generated by the computer. To be sure about it, the virtual environment was *specifically configured* to be sensible. The "laws" operating within the simulation, which determined the behavior of the all-important virtual rays of light, were not haphazard; they did not arise through accident, nor did they exist through fluke or some other random process. Rather they were deliberately contrived by human intelligence (a clear case of intelligence creating an intelligible system). And it was precisely because the virtual rays of light manifested in a sensibly ordered way that sense was able to be made of them. This sense was made through the focusing of those virtual light rays. Remember, virtual rays of light will, according to the software laws underlying them, behave in a sensible and highly ordered way—just as real light behaves in a sensible and ordered way. Once you have matter or energy or information behaving, or flowing, in a sensible law-abiding fashion, then learning and sense-making can take place in response to that sensible flow. Thus, the evolutionary line that led to the impressive round fish eye was nothing less than a kind of feeding process wherein the evolving lineage of fish eyes were feeding upon the sensibleness and intelligible order inherent in their environment (and getting rewarded for doing so). Each evolutionary step represented a keener ability to make or reflect such sense. In a very real way, the evolving lineage of virtual eyes *learned,* through trial and error, how to bend light to their own advantage. Even more to the point, it was the larger system of which the virtual eyes were a part that provoked and cultivated this learning trend.

In other words, then, the gradual evolution of a highly ordered round fish eye necessitated a system *specifically configured to facilitate just such an outcome.* By engineering sensible virtual laws (specific context) along with a population of entities (specific objects) that could potentially make sense of those laws so as to promote themselves, evolution was able to take hold and do its thing (a specific object–context interaction). And it was not just the virtual laws operating in the simulation that were sensibly configured. The very computer itself—the hardware running the software—was likewise reflective of specific design and preconfigured sensibleness. In this case, the "wiring under the board" was astronomical in terms of extent and complexity. It was precisely the presence of this designed computer system with its equally designed software that represented the initial sensible context and allowed sensible processes to proceed within the simulation. It is thus the case that the remarkable outcome of the evolved virtual fish eye was dependent, in the first instance, upon the orchestrating effects of human intelligence. Once human intelligence has engineered a sensible virtual environment in which a sensible and law-conforming flow of information is in operation, then entities inside that environment can tune into, learn about, and make sense of such information. Sensible non-random order in one form is required so as to promote sensible non-random order in another form. You need something smart to get something smart.

Considering these undeniable facts, we can see that this particular computer simulation unwittingly reveals how sensibly functioning virtual fish eyes can evolve only if a specifically configured context, or law-conforming system, has been engineered right from the word go. Sensible, nifty things like virtual fish eyes come not from some senseless vacuum; rather they emerge out of a system that is itself eminently sensible and nifty. Indeed, the a priori sensibly ordered context that determined the evolution of the fish eye actually extended beyond the contrived software laws of the program and the hardware of the computer, for these two sensible systems were themselves embedded in the sensible context of the science community and the culture of which this community is a part. To reiterate, there is no creativity coming out of nowhere. Ordered and organized things do not spring from non-ordered, non-organized things. Clever things do not come from non-clever things. There is no information being created ex nihilo. What we have instead is a system that must be intelligently configured in advance. Once it is in place, evolution can happen, a process whereby the inherent sensibleness of the system can be reflected. Which means that virtually evolved objects and their virtual environment are inseparable components of a *single intelligently configured informational system.*

The relationship whereby an enveloping context (i.e., a greater whole) exercises some degree of control over the behavior of its constituent parts is sometimes referred to as holistic causation or downward causation. Under the influence of holistic causation, the constituent parts of a system will be obliged, or be constrained, to behave in a certain definite way. With the evolution of the simulated fish eye it is clear that the driving force behind its evolution extended right through the sensible software and sensible hardware of the computer and into the sensible psychological domain of the scientists who configured the simulation. One can divine a kind of nested informational hierarchy whose influences spread from the greater and general whole to individual and localized parts. Although it may appear as if a virtual fish eye can evolve in isolation and without any intelligence or purpose, the greater system, or context, of which it is a part most certainly is replete with intelligence and purpose, the evolution of the virtual fish eye being a direct reflection of this intelligence and purpose. Which again means that it is the sensible and lawful configuration of an environment that drives any evolutionary process within that environment. *Once an environment has been intelligently configured it will possess sensible properties, and it is precisely these sensible properties that evolving entities respond to and learn to make sense of.*

Due to its importance, let me provide the reader with another example of the contextual principle under discussion. This time we will consider another computer simulation of evolution, only this time it will be an entirely imaginary one. Although it is purely hypothetical and extremely simplified, if the envisaged simulation were to be carefully implemented and run, then it would almost certainly do exactly as inferred (to be sure about it, evolutionary logic dictates that it would have to). The imaginary simulation goes like this: Instead of virtual eyes evolving to make sense of virtual rays of light, we'll imagine virtual entities that can "feed" upon the numbers produced by a number generator within their virtual environment—if, that is, the entities can correctly guess the numbers produced by the number generator, or can at least come numerically close. The numbers produced by the number generator can be thought of as potential food that can promote fitness. Guessing a number correctly therefore corresponds to eating a big fitness-increasing meal. Getting numerically close to a number would be akin to getting a small snack. A few numbers out and nothing is gained.

Let's say that this number generator outputs a number between 1 and 100 each and every second. We can set up the entities so that they are likewise capable of producing a number between 1 and 100 every second, although they do this a split second before the number generator does so. If one of these entities can correctly anticipate the number about to be produced by the number generator, then the entity will gain a fitness advantage. Guess a number correctly and fitness is conferred. Get very close to guessing the number and only a slight fitness advantage is gained. A few numbers out and there is no reward at all. For argument's sake, we can imagine that after one minute, or sixty guesses, the members of the initial population are measured in terms of their success (the number of "hits" or "near hits" attained) and rewarded accordingly. Those entities that are more accurate in guessing the numbers get rewarded (i.e., positively selected) by being allowed to replicate a little more successfully. The next generation—some with slight modifications so as to mimic mutation and variation—are likewise tested and so on.

Now, it is clear that if the number generator is a random number generator, then none of the entities will be able to evolve. No matter how they mutated in terms of the guessing strategies they used, none would be able to fare any better, since there is no good strategy whatsoever for guessing the outcome of a random number generator. Since no sense can be made of the number sequences produced, there will be no evolution. If all the entities start out with purely random guessing, all will fare equally well on average. And there will be no selective pressure to evolve since there is no sense to be made of the numbers (i.e., the number sequence produced by the number generator is nonsensical and devoid of any patterns or order). No sense to be made and consequently no evolutionary direction toward which to gravitate. True, one entity might correctly guess a number by accident and then be granted a fitness advantage, but it does not take a genius to realize that none of its offspring would be any the wiser. So whilst there might be some change in the "virtual gene pool" of these simulated entities over time, the change will be directionless and will merely drift hither and thither. The kind of constructive evolution with which we are familiar and that was evinced so markedly in the fish eye simulation will clearly not transpire in our imaginary simulation.

If, however, the number generator were programmed (i.e., specifically contrived by human intelligence) to output *even numbers* between

1 and 100 (in no particular sequence, though) instead of random numbers between 1 and 100, then evolutionary logic dictates that there will be an inevitable trend toward the selection of entities that have a strategy for choosing more even numbers than odd numbers. An *inherent direction* toward an *inherent target* will now be apparent in the particular way in which the entities change. As the entities mutate and vary, sooner or later we can imagine that some will happen to mutate in such a way that they choose more even numbers than usual, to be biased as it were. These variants will be rewarded in terms of reproductive success and will eventually come to dominate the population. What the successful variants will have done in this case is to have gradually begun to make sense of the numbers being produced by the number generator. They will have begun to hone in on the right law to reflect, so to speak—namely to have a preference for picking even numbers. Moreover, given enough time an entity will evolve that always opts for even number guessing.

Once this even-number-picking strategy has been alighted upon through cumulative selection, then it will represent the maximum amount of evolution that can be achieved by the simulation. Clearly with a non-random number generator in operation, there is potential sense to be made, and this is exactly what any evolutionary process does—it is a process in which sense is made, or reflected (hence my repeated contention that evolution involves *sensible changes* to a heritable code and not just any old changes). In this case, an entity will eventually evolve that can make optimum sense of the number generator. It will achieve this by always picking even numbers between 1 and 100. The instructions for making optimum sense will be encoded in the entity's underlying computer code.

In order to elicit more elaborate forms of evolution in this hypothetical simulation, we would need to make the number generator a bit more complex. For instance, we could contrive the number generator to produce the sequence 2, 4, 8, 16, 32, and 64, repeatedly. Evolution would then invariably steer the entities into coming up with strategies that are wise to the specific rule or law underlying this sequence, or at least wise enough. We should first expect the entities to evolve toward picking even numbers only. Then we might expect them to evolve a strategy in which a progressive sequence of even numbers is chosen. Finally, after untold generations of fitness testing, we would see a strategy emerge that effectively matches the actual law underlying the repeating number sequence.

The point of all this is to show that evolution happens only if there is sense to be made, and sense can be made only if it is configured into the greater system within which evolution occurs. Once a system has been intelligently configured with sensible rules and laws (i.e., an intelligently configured context is set up in advance), then evolution can happen within the system in response to that initial sensibleness. In a real way, evolutionary processes are part of a stimulus/response routine. The stimulus lies with an ordered and sensible context that drives a specific evolutionary response.

We shall learn more about simulated forms of evolution in the following chapter, since they are so very illuminating with regard to contextual principles and lend themselves to the natural intelligence paradigm. For now, it is enough to realize that sensible evolutionary changes happen only amid an already sensible system. Without the primary ingredient of a smartly configured context—of which all-pervasive laws and orderly rules are the chief features—there can be no evolutionary events whatsoever. In the last analysis, lawful and sensibly configured contexts are absolutely necessary for evolutionary processes to proceed. Making sense is the name of the evolution game.

CONTEXT IN THE REAL WORLD

Having covered the virtual fish eye, our other example under consideration is the evolution of complex structures within the natural world—like our tiger or even real eyes like our own. In this case, it is Nature itself, with its various laws, that acts as the sensible context or sensible environment that allows biological evolution to proceed.

By highlighting particular variations of DNA, by putting a spotlight on them and testing them so to speak, Nature will select and thence hone those changes in DNA that make the most sense in terms of the biological behavior that they code for (this is especially the case with key biological processes like protein synthesis). This again drives home the point that evolution involves not simply genetic change but rather sensible genetic change. It is this inevitable trend for bio-logic to make more and more sense that imbues evolution with an inherent direction and an inherent target—a blatant fact that is denied by the majority of evolutionary biologists.

All this suggests that the more sense and intelligibility inherent in a system, the greater the amount of sense that can become reflected by an evolving entity within that system. The greater the a priori sensibleness in a world (real or artificial), the greater and more complex can the evolution be within such a world. Put another way, the more intelligently a system is configured in terms of its laws and its rules, the more intelligence can be generated by an evolutionary process within that system; the more intelligent the contextual stimulus, the more intelligent the evolutionary response. (I have a suspicion here that such reasoning is open to an informational analysis—that one could restate the foregoing in terms of information theory—but this remains to be seen.)

Thus, considered as a system, Nature is surely the ultimate sensible environment in terms of the biological evolution and biological innovation that it has inspired. Life can be seen as an ongoing process in which more and more of Nature's abundant sensibleness is being tuned in to and being made sense of. This is particularly evident in the evolution of brains and nervous systems. The more evolved a brain or nervous system is, the more able it is to make sense of the various fields of information in which it is immersed. What feeds and fuels human intelligence is Nature's sensibleness. If Nature were not rich in intelligibility and sensibleness, then there could be no human intelligence, for there would be nothing to learn, no sense to be made, nothing to figure out. In light of this, I submit that *intelligence and intelligibility are two sides of one coin.* The only reason Nature is intelligible is because Nature is intelligently configured—just as the only reason a virtual computer environment is intelligible is because it is intelligently configured. Thus *intelligence begets intelligibility begets intelligence.* Nothing, it seems, could be plainer and yet simultaneously so controversial as those five words.

As the reader might suspect, the issues under scrutiny here are rife with subtlety and import. So at this point in our tentative apperception of natural intelligence, let us inquire more into the role of sensible environments in eliciting evolutionary processes. We can start by considering further what computer simulations tell us about intelligently configured contexts and their necessity if evolution is to manifest and "do its thing." As we shall see in much more detail, artificial forms of evolution substantiated within a computer simulation all demand an initial input of intelligence, this intelligence being the force that sets up the software laws (and the hardware, of course) that will determine the evolutionary potential of any simulation. It is only after this input of intelligence has been brought to bear that evolutionary processes can subsequently take hold. The same can be said, of course, of Nature, the difference being that Nature represents the ultimate intelligent system according to which the evolution of the real tree of life has been elicited. In a nutshell, evolution is a naturally intelligent learning process provoked and cultivated by a naturally intelligent Nature. To fully grasp this principle of intelligence is to embark upon a conceptual revolution.

CHAPTER 4

BINARY ACORNS

The Science of Artificial Life

Many middle-aged readers will remember with a mixture of nostalgia and embarrassment the emergence in the late 1970s of Space Invaders, a cult computer arcade game in which one had to fight tooth and nail against a constant stream of invading aliens. In its day this joystick-gripping game was state of the art in terms of computer gaming and computing power. I even used to bunk off college in order to go and play it. There were mesmerizing screens of digital invaders, their assaults relentless and always ending in victory. These attacking aliens, made of only a handful of luminous green pixels, certainly seemed alive and intent on destruction as they crept menacingly across the computer screen. But, of course, the diminutive throbbing ETs were not really alive, for they could not genuinely replicate (even though some of the critters would divide into two when one blasted them), nor could they evolve. The game, or computation, was totally preprogrammed and pre-dictable. In contrast, the computations involved in the contemporary science of artificial life are in a different league. The alienesque entities created by artificial-life scientists are much more lifelike, so much so that some claim them to be really alive.

At first glance, the study of artificial life sounds suspiciously like the study of fake life. Are the scientists who create and pursue artificial life forms within computers therefore forgers of some kind? Well, yes and no. Yes, because it could be argued that the digital entities that they study within computer environments cannot be real living things despite any claims to the contrary. Certainly these digital entities are not made of flesh and blood. Such digital organisms do not smell or bite or yelp. Nor do they breathe or eat. And unless transported on some sort of disk or launched on the Internet, they cannot run away. Yet they can behave in ways that are distinctly lifelike. Which is to say that they can exhibit behavior that is akin to eating and biting. Moreover, unlike preprogrammed immutable Space Invaders, modern artificial digital entities can be subject to evolution, which can lead to the emergence of various forms of behavior that bear the most distinctive of life's hallmarks, namely the ability to make acute sense of the environment and the ability to act upon that sense (and hence reproduce).

In short, the science of artificial life can teach us much about the fundamental principles involved with evolution, and this is its chief virtue. More importantly, this new science also makes it clear that simulations of evolution all require an initial input of intelligence if they are to succeed in yielding meaningful and significant phenomena. It is the unambiguous presence of a sensibly ordered contextual system that betrays the intelligence (in this case human intelligence) operating in the evolution of lifelike entities. Without intelligent configuration of one kind or another there can be no evolution and, in consequence, no cleverly functioning entities. Human intelligence engineers smart computer systems, and, in turn, these smart computer systems elicit smartly behaving virtual entities. Always and everywhere, intelligence is the prime creative mover.

LIFE ON THE SCREEN

The study of artificial life has really taken off in the last few decades. A branch of artificial intelligence, artificial life is now a respectable science in its own right. As computers get faster and faster, more and more rapid forms of simulated evolution can be run on them. As noted, many experts in evolutionary biology see in these computer simulations further evidence that evolution, in terms of its defining characteristics, is an essentially dumb and mindless affair. But this interpretation is applicable only if we ignore the presence of sensible contexts, which, in the case of computerized forms of evolution, are the designed software rules running the computations: the designed virtual environment, in other words (note also that this design extends out into the hardware "environment" within which the software is embedded). And how often have we heard how incumbent upon us it is to consider the environment? It seems then that the acknowledgment of the role and status of the environment is important in many domains.

Before we return to this strangely contemporary issue of the environment and how we often neglect to think about it (especially when thinking about evolution), we shall first take a look at the work of artificial life researchers like T. S. Ray, an ecologist, biologist, and inventor of a pioneering artificial world called Tierra. As we saw in the previous chapter with the virtual eye experiment, computer simulations of life processes start off with a population of digital entities that exist solely inside a computer system. T. S. Ray, our main informant, puts it like this. The computer system becomes home to

> a population of data structures, with each instance of the data structure corresponding to a single entity. These structures contain variables defining the state of an individual. Rules are defined as to how the individuals interact with one another and with the environment. As the simulation runs, populations of these data structures interact according to local rules, and the global behavior of the system emerges from those interactions.[1]

As to the medium, contextual space, or virtual universe in which this population of digital entities exist, Ray explains:

The computational medium of the digital computer is an informational universe of boolean logic, not a material one. Digital organisms live in the memory of the computer and are powered by the activity of the central processing unit (CPU). Whether the hardware of the CPU and memory is built of silicon chips, vacuum tubes, magnetic cores, or mechanical switches is irrelevant to the digital organism. Digital organisms should be able to take on the same form in any computational hardware and in this sense are "portable" across hardware.[2]

Ray goes on to explain the laws that govern the evolution of digital organisms. Whereas living organisms are subject to things like the laws of chemistry and physics, digital organisms

might as well live in a different universe from us, because they are not subject to the same laws of physics and chemistry. *They are subject to the "physics and chemistry" of the rules governing the manipulation of bits and bytes within the computer's memory and CPU.* They never "see" the actual material from which the computer is constructed, they see only the logic and rules of the CPU and the operating system. These rules are the only "natural laws" that govern their behavior. They are not influenced by the natural laws that govern the material universe.[3] [my italics]

As we shall see, it is precisely these intelligently written rules and laws that represent the sensibly ordered context that draws out, or facilitates, the subsequent digital action. But for now we need only to grasp what artificial life experimentation is all about, namely, that a group of data structures (basic programs consisting of strings of binary code) are introduced into a virtual environment where their evolutionary fate is determined by software laws—"the logic and rules of the CPU," as Ray puts it.

The next step is to have the digital organisms compete over some resource and also to replicate themselves. To fully simulate evolution it is necessary that replication be dependent upon some form of measurable success. Recall that with our earlier virtual eye experiment, success was measured by the ability to focus virtual rays of light. And with the number generator simulation I asked you to imagine, success was measured in terms of the ability of the entities to anticipate the numbers produced by the number generator. For Ray's artificial organisms, success can mean the ability to utilize computer memory (RAM) and CPU time, which can both be considered as basic exploitable environmental resources (akin, perhaps, to high-energy sunshine). These resources, these properties of the virtual environment, are logical, and it is this intelligible logic or sense (i.e., meaning) that the digital organisms can potentially become wise to and exploit. If one also adds random variation and random mutation (known as "bit flips") into the mix, then one is ready to instigate evolution:

> Bit flips may be introduced at random anywhere in memory, where they may or may not hit memory actually occupied by digital organisms. This could be thought of as analogous to cosmic rays falling at random and disturbing molecules that may or may not be biological in nature. Bit flips may also be introduced when information is copied in the memory, which could be analogous to the replication errors of DNA. Alternatively, bit flips could be introduced in memory as it is accessed. This could be thought of as damage due to "wear and tear."[4]

Thus bit flips (discrete changes in the on or off state of bits) act as a mutagenic force causing variation. Armed with such cunning software programming, you are ready to run an evolutionary simulation. Generate an initial set of digital organisms, vary them through simulated mutation, then measure their fitness according to how well they can exploit, or make sense of, the resources extant in their virtual world. Then let the digital organisms reproduce according to their fitness. Then do the same again. And again. Millions of times. Let the computer laboriously crunch out the reiteration for you. Generate, test, generate. Generate, test, generate . . .

What happens? Well, the reader already knows the gist of it. It transpires that if you let the process run for long enough, eventually the digital organisms will attune themselves to the sensible order of their environment. The digital structure of these virtual organisms—which embodies what they do and how they behave—will inevitably evolve in a way that ensures that they gravitate toward making optimum sense of their environment, which, in this case, consists of the laws and such that operate within the computation, as well as the other data structures with whom they are competing. This is classic evolution in action. Moreover, it works. Digital entities have a real existence, and they can evolve over time just like living organisms. As artifical life researcher Robert Pennock avidly attests: "All the core parts of the Darwinian process are there. These things replicate, they mutate, they are competing with one another. The very process of natural selection is happening there. If that's central to the definition of life, then these things count."[5]

To drive the message more forcibly home, it should be realized that genetic algorithms are now big business and are used in industry. They have been employed with great effect in areas as diverse as airport logistics, medical diagnostics, jet engine design, factory optimization, and resource management. Imagine you need a computer program to carry out some highly complex function, a program that a lone human programmer could not possibly hope to write, like how to manage landings and takeoffs at an airport subject to sudden and unexpected delays. The best way to obtain such a program might be to quite literally evolve it by utilizing a genetic algorithm. As long as you have an initial population of programs that in some way, no matter how small, attempt to address the problems at hand, then, through selection according to measurable success, an optimal solution can be converged upon.

Steven Levy, author of *Artificial Life,* puts any skepticism we might have about all this in the following way:

It seemed an almost absurdly simple recipe for optimization: take a string of random numbers and treat them as computer programs. Grade them according to how well they do at executing the work of a custom-designed computer program, and then reward them to the extent of their excellence by allowing them to reproduce to that degree. Then take the revised population, pair the strings, and have each marriage partner swap a part of itself with its mate. Change a few bits for mutation, and do it again. One would intuitively expect this process to take a very long time to match the results of a computer program specially written for a task—in fact, it might be difficult to envision something that good *ever* resulting from this elementary process.[6]

But, as Levy goes on to point out, good results do happen. In fact, given the right sort of virtual environment, genetic algorithms are *bound* to work. It's an unquestionable form of self-organizing logic that, once substantiated within some system—in this case a computer system—will simply have to yield creative results. If you take the head-scratching problem of how to get a computer system to recognize, say, people solely of Chinese descent (as opposed to other races who may have similar facial features), then you can potentially breed a solution within a computer system. You take your initial population of attempted "Chinese spotters" (i.e., software programs all slightly different from one another that are "exposed" to various faces), measure the success of each, and then select those programs whose ability to discriminate faces is slightly better. Breed from these and make minuscule alterations to the next generation. And repeat ad infinitum. Eventually, after perhaps millions of generations, the genetic algorithm will yield an optimum solution. If there is any sense in the data that the population of programs is attempting to determine (i.e., if there is genuinely some sensible pattern that defines a Chinese face), then you can be sure that gradually, bit by bit, a species of program will arise that makes sense of that unique pattern, thereby ensuring its own survival. Of course, this assumes that there *is* a law-abiding pattern to be divined (I am sure that different races of people do indeed have facial characteristics that can be formally described in terms of various mathematical ratios and such). In any case, the point is that smart programs can be evolved using genetic algorithms. Which means that evolution is a simply great way of creating smart things and smart behavior. As a type of process, evolution works every time.

SOMETHING OUT OF NOTHING?

For many observers, it once again appears as if simulated evolution, just like real-life evolution, is a method of creating clever and meaningful things out of nothing, the only difference between artificial evolution and real-life evolution being the time needed to achieve results. Echoing the sentiments of others, Levy states:

> [Genetic algorithms] have often been attacked on the grounds that natural evolution is simply too slow to accomplish anything useful in an artificial learning system; three billion years is longer than most people care to wait for a solution to a problem. But computer muscle telescoped millions of generations worth of evolution into a lunch hour, and the GA [genetic algorithm] turned out to be a stunningly powerful tool. Indeed, it seemed to deliver . . . "perpetual novelty" and "something out of nothing."[7]

This is simply not the case. Something cannot come from nothing. To evolve an ordered and sensible string of computer code or DNA code, such code must be subject to the scrutiny and informative influence of a sensibly ordered environment. There has to be sense and meaning of one kind to generate sense and meaning of another kind. Or, to put it another way, there must be specifically organized information of one kind to generate specifically organized information of another kind. Take the example of the Chinese face recognition genetic algorithm.

An optimum solution can be evolved only if there is genuinely a sensible pattern underlying Chinese faces (caused perhaps by a certain invariant constellation of genes). This implies that there is a subtle order, an intelligible law-abiding pattern of some kind, that can be made sense of.

If, however, Chinese faces are devoid of any genuine common pattern, then no matter how long one ran a genetic algorithm, no optimum solution for Chinese face recognition would be found because none exists (the genetic algorithm would be unable, say, to differentiate between Chinese faces and Japanese faces). On the other hand, if we could indeed evolve such a face recognition program, then it could potentially be further evolved so as to be able to read facial expressions (i.e., recognize happiness or sadness according to facial shape) as long as the software was programmed to reward instances of correct recognition and so long as facial expressions do indeed obey certain laws (which they appear to do, since smiling and frowning are universal). Here again we can see that for a genetic algorithm to succeed in producing a really smart piece of software, one needs some sort of a priori sense and order, which evolution can tune in to and home in on. Without a priori sense, without an a priori pattern, without a priori contextual order, there can be no evolution. However, reductionism or merelyism will not alight on this insight, since focusing attention solely and exclusively upon the isolated parts of a system reveals little about context.

The bottom line then is that the digital entities evolving within a computer simulation cannot be understood solely in terms of themselves alone. Their fate, their apparent "dumb" evolution into smarter and smarter forms, is entirely dependent upon the environmental context in which they find themselves. Only if an environment is sensible can sense be made of it. Only if an environment is graced with robust and smartly configured laws can smart lawful events unfold. Only if an environment is replete with specific patterns of information can other forms of specific information arise therein. In essence, a virtual environment or virtual world running on a computer system must be infused with intelligibility if creative evolutionary processes are to emerge. It is this intelligibility of a computational environment that reflects the original (human) intelligence that allows artificial evolution to proceed (recall my contention that intelligibility and intelligence are two sides of one coin). Even if a simulated form of evolution yields entities (i.e., programs) whose powers and capabilities far exceed our wildest expectations, it is still the case that the evolutionary magic was instigated by human intelligence, which set up the computation and implemented a specific flow of information within it. One absolutely cannot get something from nothing, and it is flabby reasoning to suggest otherwise. The intelligently behaving products of an evolutionary simulation demand an intelligently configured context according to which their arising is assured.

The same principle must prevail in Nature. The entire system of Nature—its given configuration and what this configuration achieves—is patently not dumb and mindless but an embodied expression of natural intelligence, the intelligence being distributed and manifest throughout the system, in both the laws of Nature and the constructive material stuff of which Nature consists. Meaning begets meaning, order begets order, and intelligence begets intelligence. Biological evolution is driven by the intelligence inherent within the system of Nature, just as artificial evolution is driven by the intelligence inherent within a computer system. In whatever medium it manifests, evolution is an intelligent response to an intelligent stimulus.

Interestingly, a number of intelligent design creationists, such as William Dembski, have criticized the evolutionary prowess of genetic algorithms on the grounds that they cannot generate complex smart programs without first having an input of intelligence. Dembski argues that even though a genetic algorithm might produce something clever, it is only because human intelligence coordinated the computer system in highly specific ways to begin with. This is correct, of course—but the wrong conclusion is drawn with respect to real-world evolution. Rather than seeing intelligence as being part and parcel of the larger system of Nature, the only place Dembski can divine the intelligence is in the form of some mysterious supernatural Designer lying outside of Nature. For some reason that is hard to fathom, hard-core creationists simply refuse to reinterpret life and evolution. If they did, if they saw evolution as a naturally intelligent process initiated by a naturally intelligent Nature, and if they saw life—*all life,* not just spectacular instances of life such as bombardier beetles or bacterial proton rotors—as a naturally intelligent technology, then they would not need to invoke supernatural entities (which themselves would require even more explanation!). Yet all parties—the creationists and most of the scientists with whom they argue—steadfastly refuse to see that self-organizing intelligence is a fundamental property of the Universe and its laws and its Lego-like components. One does not need to invoke something extra on top of all this. For some people, however, even the whole of Nature is not enough intellectual or spiritual nourishment.

NATURE'S INHERENT INTELLIGENCE

The upshot of all this, the chief lesson, is that evolution in the real world by way of DNA and genes is possible only because of the remarkable law and order inherent within the totality of Nature, a form of law and order, which facilitates the very arising of DNA, as well as the subsequent organismic magic that DNA can potentiate. Without such a sensibly ordered environmental context, DNA code would not be able to express any sense whatsoever. In fact, *nothing would make sense,* for sense is utterly context-dependent. Thus, it is the innate sensibleness of Nature—what Einstein called the mysterious comprehensibility of the Universe—that drives the evolution of all life. If we conceive of evolution as a sense-making process (all organisms can be considered as sensible systems of bio-logic), then clearly there must be sense to be made. Nature and its laws are the source of all this ambient sense, evolution being a *localized reflection or translation of this sense into a new form.*

The fact that life has tuned in to the laws of chemistry evinces this principle. As a very simple example, consider the cement that barnacles employ, which allows them to adhere to rocks in the face of powerful tidal forces. Incredibly, a layer of this cement three ten-thousandths of an inch thick over one square inch will support a weight of seven thousand pounds. That in itself is an amazing example of the kind of engineering acumen of barnacle bio-logic. But upon what, exactly, is the acumen built? The glue is a viscoelastic gel made from various proteins (and water). Barnacle genes are what code for the specific proteins in question. The genes involved thus contain the information required to make the right sorts of complex macromolecules that can form into extremely strong cement. Which means that the barnacle has tuned in to the sensibleness of chemistry and has reflected part of that sensibleness within itself. In other words, the acumen, or smartness, evinced by this example of barnacle bio-logic is driven/informed by Nature's chemical laws. This applies to *all* the protein manipulation occurring within the barnacle as well as in all other creatures. Life has learned the art of chemistry, the art of manipulating chemicals and molecules in sensible life-enhancing ways. If there were no sensible chemical laws operating in the environmental context we call Nature, then there could be no learning and no life. One can even envisage molecular chemistry as being a language of Nature and that life has learned to make sensible "utterances" with that language. In a very real way, the whole of Nature serves as a *tutorial context* that teaches life to "speak" sensibly, or at least encourages life to author its DNA in a clever and meaningful way.

With this kind of reasoning it is apparent that an environment is not some minor incidental aside. Rather, an environment represents a richly meaningful context according to which processes like evolution are galvanized into constructive action. It follows that the greater the inherent sense and order within an environment (the more smartly it is lawfully configured), the greater the fruits of any evolutionary process will be within that environment. The more sensibleness there is within an environment, the more a learning process (like evolution) can be facilitated.

In the case of the evolving virtual fish eye mentioned earlier, the eye will actually reach an optimum state in which no better focusing of virtual light rays can be achieved. After all, the contextual environment it finds itself in is severely limited in terms of the sensibleness available to be made sense of. No matter how many times the computation is run, a certain optimum fish eye will *always* be reached after which *no further evolution can take place.* Unless, of course, you manipulated the virtual environment and coded for some more sensible variables. If you did this, if you provided more lawful sense within that environment by, say, introducing some sort of specific sensible object that the virtual eye could potentially "see" in order to gain some fitness advantage, then more evolution could be fostered (especially if the digital code underlying the virtual eye could be indefinitely extended in length). Introduce some more sensible and lawful parameters and evolution can "reach out" in new sense-making directions. We saw this exact same principle in operation in the hypothetical computer simulation in which evolution would cease once the virtual entities had made optimum sense of the rules or laws determining the number generator's output. Only if the rules or laws governing the number generator's output were modified in some clever way could we get evolution to respond in more elaborate ways. To reiterate: the more intelligently any given system, or world, is configured, the more intelligence can be generated within it. Indeed, this notion is well understood by virtual life researchers, as their simulations often reach an evolutionary impasse. Only by making a virtual environment richer in information and sensibleness can one stimulate more progressive forms of evolution.

In the real world, nervous systems can likewise become more and more evolved, more and more sensitive, only if the world around them provides a wealth of sensible data that can be made sense of. Picture in your mind's eye the evolution, say, of the mammalian brain from a shrewlike ancestral form right up to the present human incarnation. What is happening over time is the gradual complexification of the brain—a transition from Brain 1.0 to Brain 12.0, as it were. As a brain evolves in complexity, so, too, does its capacity to process and organize information. The more evolved the brain, the more sense it can make. Where does this sense that the brain feeds upon come from? The answer is Nature—the natural world within which organisms exist and that is replete with tutorial order on all scales. Our brains perceive and process this ambient order via our senses, which, as their name implies, serve to channel the informational stuff of which sense is made. It is this inherent sensibleness of Nature (on all scales, I might add, from the organized world of atoms and molecules to the organization of galaxies) that is so indicative of intelligent configuration.

Take another example. Quite apart from their inner activity, the pointed shape and waxy surface of most tropical plant leaves pays testimony to the intelligible and sensible configuration of the world around them. These properties allow rainwater to readily drain off the leaf's surface, thereby allowing photosynthesis to proceed unhindered. The reason water will drain off in a predictable and ordered way is due to its precise molecular properties and the precise law of gravity to which it adheres. It is these kinds of sensible features of the natural world that have encouraged specific leaf shapes/surfaces to evolve. Were the movements of water and the rest of the natural world capricious and fickle, there could be no evolution of sensible biological structures. The same is true with the inevitable evaporation of water when a certain temperature is met—this is a phenomenon that mammalian bio-logic has made brilliant sense of so as to foster temperature control (via sweating). Similarly, with metabolism—probably the most essential of all life's innovations—bio-logic has learned to tune in to the laws of chemistry and can thereby manipulate energy through the enzymatic control of certain chemical reactions. When one considers that all these examples represent no more than a *negligible fraction* of bio-logic's capacity to reflect and embody sense, one can conceive that the natural intelligence of bio-logic *must* reflect the natural intelligence permeating its contextual surround, an intelligence that ultimately resides in Nature's laws and the extensive order induced by these laws.

The tree of life, in all its eclectic organismic aspects, is therefore sown, grown, and nourished according to the intelligence enveloping it.

This "enveloping" represents the sensible laws of Nature and the various physical and chemical effects of these laws upon gross matter. It is in accordance to this sensible and "stimulating" context that the tree of life attunes itself and evolves. This is not unlike the situation whereby we tune in to an informative radio broadcast. As we turn the dial and tune in more accurately, the broadcast becomes clearer and we can then follow its inherent sense. Evolving bio-logic does the same kind of thing, only in this case the sensible broadcast being tuned in to is the rest of Nature.

COEVOLUTION

To further highlight the inevitable success of an evolutionary process to produce smart and sensible structures and the role of context in ensuring this, consider another example of artificial evolution. This one is a little more elaborate and involves *coevolution,* wherein two sorts of artificial organisms compete with one another, thereby stimulating each other's evolutionary development. In the real world such coevolution happens with parasites and their hosts (symbiosis, as we shall see in a later chapter, is also a form of coevolution). Parasites, by definition, are organisms that live off other organisms for their own benefit and not to the benefit of the host (like fleas, leeches, and ticks, for instance). Realizing that parasites can push their hosts into evolving defensive strategies, experts in genetic algorithms have tried to foster the dynamic pairing of artificial parasites with artificial hosts in order to enhance the evolutionary progress of their simulations.

According to Kevin Kelly in his classic book *Out of Control,* it is in this sort of coevolutionary domain that the future of genetic algorithms might lie. In discussing the work of computer designer Danny Hillis, Kelly writes:

Hillis proposes setting up a swarm system which would try to evolve better software to steer a plane, while tiny parasitic programs would try to crash it. As his experiments have shown, parasites encourage a faster convergence to an error-free, robust software navigation program.[8]

This sounds strange, even disturbing, especially if we imagine that one day the results of this sort of research will remove the need for human pilots. Many of us would undoubtedly make alternative travel arrangements rather than put our lives in the hands of some evolved algorithm-based pilot that is passed around on a memory stick for inspection before a commercial flight. Joking aside, what Hillis is suggesting is that in order to generate a really, really smart computer program that can safely fly a plane (an obviously complex task), the program needs to have evolved in a really harsh and consistently varying environment. Not a wholly random environment, mind you, but an environment that is not always easily predictable. The presence of parasitic programs attempting to foil a population of pilot programs is a way of ensuring that the pilot programs are pushed into evolving more and more flexibility. The parasitic programs can be likened to adverse weather conditions such as turbulence, which, although not absolutely random, can vary enough to ensure that any competent pilot program needs to be distinctly plastic and adaptive. Smarter, in other words.

Artificial parasitic organisms thus act as a further selective pressure in artificial evolution. In turn, this implies that already-established species of artificial life are themselves a source of sensibleness, and that for artificial organisms to further evolve depends upon them making sense of their *entire environment.* It should now be becoming clear just what the term "environment" really means. Whether the real world of Nature or the artificial world of computer simulations, the environment ultimately consists of *everything,* both already-established organisms and the contextual forces (laws) and elements around them.

Such systems, such worlds, are infused with intelligence, the intelligence residing in the lawfully configured context of the system. This "primary" intelligence then becomes quite literally reflected in both the evolutionary processes fostered within those worlds and the actual organisms wrought by those evolutionary processes. Indeed, living organisms can be considered the established *records* of such a reflectional or translational process, the records written, of course, in DNA and able, en masse, to catalyze further forms of evolution. To survive, organisms need to make sense of *all* their surroundings, including not just the physical and chemical law reigning in their surroundings but also the sensible construction and behaviors of their fellow species. Natural intelligence can accelerate, becoming reflected in quicker and quicker durations, because evolved organisms feed back on one another and stimulate further evolutionary developments. As a sense-making process, evolution is rather like a tail-eating serpent feeding off itself, off its past handiwork if you like, thereby allowing it to establish itself more and more concretely. As intelligence becomes reflected within biological systems (i.e., bio-logic evolves to become wise to physical and chemical law), extant biological systems act as further stimulus for evolution.

NATURAL INTELLIGENCE VERSUS CHANCE

The problem, or shortfall, of evolutionary dogma should now be all the more clear after our foray into simulated forms of evolution. Evolutionary biologists are wont to focus the bulk of their attention upon the chance events operating within evolution, most notably mutations and variation. By concentrating attention on these random components and, in consequence, by elevating their level of importance, evolutionary science ignores a far more salient aspect of evolution, namely the contextual features of an environment that serve to facilitate and encourage evolution As I have been at pains to point out, these contextual features are both intelligible and indicative of intelligence. Without an intelligently configured context "surrounding" a mutant or variant life-form, whether fleshy or digital, there can be no selection process. For successful mutations to be selected, they have to make some kind of sense and have some kind of sensible life-affirming effect—so something must say, as it were, "Yes, you have correctly made sense of me." It is precisely an intelligently configured context that affords this kind of tutorial interaction. Smart contexts draw out smart evolutionary processes. The tree of life is smart because its surroundings are smart.

It is imperative at this juncture that an understanding of this issue of context be firmly grasped. I labor the point since paradigm shifts initially require persistent force. Indeed, the academic world is so entrenched in neo-Darwinian dogma that anyone who refuses to walk the orthodox line is invariably treated as a miscreant and curtly dismissed. To state that intelligence is a property of the whole of Nature is tantamount to blasphemy to many scientists. Anyhow, consider the following quote from esteemed scientist Jacques Monod, a quote that captures the old evolution paradigm in all of its misguided glory. The quote is taken from Monod's famous book *Chance and Necessity,* which asserts that biological evolution is essentially a chance affair. As we shall see, Monod's mistake is that he takes things out of context (and we all know how misleading it is to take things out of context!). Monod sums up his merelyism as follows:

> Various mutations have been identified as due to: (1) the substitution of a single pair of nucleotides for another pair; (2) the deletion or addition of one or several pairs of nucleotides; and (3) various kinds of "scrambling" of the genetic text by inversion, duplication, displacement, or fusion of more-or-less extended segments. We say that these events are accidental, due to chance. And since they constitute the *only* possible source of modifications in the genetic text, itself the *sole* repository of the organism's hereditary structures, it necessarily follows that chance *alone* is at the source of every innovation, of all creation in the biosphere.[9]

Monod is not wrong in the first part of his statement. Obviously mutations—changes in genetic code—are necessary if evolution is to take place. This is true as much for the evolution of virtual life-forms as it is for flesh-and-blood life-forms. Where Monod errs is in his out-of-context conclusion. As we have established, for Monod and most other evolutionary scientists, no one seems to pose the question as to *why* any particular mutation will be selected. Or, if they do think about it, they might simply conclude that a mutation is selected if it confers fitness. *But the thinking seems to stop there and is not explored in more depth.* If evolutionary scientists were to expand the focus of their mind's eye and attend to the bigger picture, they would soon realize that context is far more important than mutations in fostering evolutionary creativity and evolutionary innovation. Context is everything.

In order to help lay Monod's dull and myopic sentiments to rest, let's think of another hypothetical evolutionary scenario. Think of a population of ocean-dwelling bacteria. Let's say that they fare better if they manage to consume a certain chemical within the sea, that whenever they absorb this chemical they become fitter and therefore prosper. Perhaps this chemical gives the bacteria an energy boost, which allows them to metabolize, move, and reproduce more quickly (i.e., the bacteria can make life-affirming sense of the chemical). Let's imagine that the useful chemical in question is produced by vents on the ocean floor that spew out molten lava. Produced through the reaction of molten lava with seawater, this chemical then diffuses into the surrounding area: the farther away from the vent, the lesser the concentration of the chemical.

Now, let us suppose that our population of bacteria absorb this chemical whenever they happen to move through it. Obviously, if the bacteria could search for the chemical rather than leaving its acquisition in the hands of happenstance, they would be at an advantage. Therefore, let us conceive of a mutation within a single bacterium, a mutation that allows the mutant to *actively* locate the chemical. This could be achieved by making use of the chemical's *gradient*. By actively following the chemical's gradient, our mutant bacterium would be able to move toward the source of the chemical and thus absorb more of it (actually, chemical gradients are exploited by ants and other insects, as we shall see in chapter 7). Let us also assume that this species of bacteria already employs a temperature-sensitivity mechanism that will stop our mutant from venturing into seawater whose temperature is so high as to destroy it. So, what will happen to our "lucky" mutant?

By actively following a chemical gradient, our envisaged mutant will move toward higher and higher concentrations of the chemical and, by absorbing the chemical, will be rewarded in terms of fitness. And its sensitivity to temperature will stop it from getting too close to the hydrothermal vents that produce the chemical. With this rather decisive life-affirming advantage, our mutant bacterium will flourish, and its slightly amended genetic legacy will spread.

According to Monod, this new bacterial behavior, despite being a decidedly clever innovation, is nonetheless due solely to chance. Indeed, that is exactly how Monod concludes his statement—and most biologists would concur with him. But we have to ask ourselves, *why* was this chance mutation selected? It is clear that the reason this particular genetic change was selected was because its effects upon the bacterium's behavior made good sense. By being able to make *more sense* of its environment, the bacterium's overall fitness was increased. But why was it able to make more sense of its environment? The answer—and this is the really crucial bit—is that both the chemical and its gradient were *themselves inherently sensible.* A chemical gradient radiating from a hydrothermal vent or any other concentrated source represents a law-abiding intelligible pattern that can potentially be made sense of. Although the diffusion of the chemical from its source might be the result of the random collisions of atoms, nonetheless a *non-random pattern* of distribution emerges. Such an ordered chemical pattern, or gradient, is, through physical and chemical law, bound to emerge. That is the beauty of Nature's laws—they force specific order to arise. So any bacterium able to become wise to the sensible non-random gradient of a potentially useful chemical is going to fare well and be selected. It will have attuned itself to yet another aspect of natural law. Further, the molecular properties of the chemical in question were likewise sensible, in terms of the chemical's specific behavior with other chemicals. A chemical can be used to increase biological fitness only because the chemical's sensible properties are being tuned in to (as happens during metabolism).

In essence, then, the bacterium in question will, through a modification of its genome, have latched on to certain sensibly ordered features of its environment (i.e., it will have become de facto *wise* to the various non-random effects of natural laws). When an organism manifests this kind of wise behavior, part of the sensibleness of an environment has essentially become reflected within the organism. If there were no laws and no sense (i.e., just nonsense) surrounding an organism, it would neither be able to live nor evolve. Evolving organisms and their environment should therefore be considered holistically as aspects of *one single system in which sensibly ordered information is everywhere flowing and reflecting itself.* And what is the chief reason that sensibly ordered information is flowing and reflecting itself? The answer is the laws of Nature, laws that may sound simple but are, in fact, an expression of intelligence. Why so? Because these laws give rise to intelligibility, and it is precisely this intelligibility that feeds intelligence. Human intelligence, for example, exists only because it is constantly *informed* by the rest of Nature. We are smart only because there is so much to learn. We are clever only because we have learned things from the world around us. We are intelligent only because the Universe represents the ultimate tutorial book. Perceiving Nature as a vast system of intelligence is simply to admit that the source of the ingenious human mind is something far more ingenious.

I doubt that Monod thought too much about all this—hence his focus upon random changes to the genetic text rather than the contextual forces that serve to preserve specifically innovative changes. It's a bit like studying a rising uncoiling snake and trying to explain it without detailing the clever snake charmer who is evoking and summoning the snake's movements.

Consider again the metabolism of cells, bacterial or otherwise. Metabolism consists of the sum total of a cell's unceasing manipulation of chemicals and chemical energy whereby the cell is able to maintain its autopoietic mode of being. Metabolism is perhaps the most important technique wielded by the tree of life. One has only to think of how utterly crucial energy production is in the world economy to realize that it must be of equal importance within living creatures. And clever ways of producing and manipulating chemical energy can be achieved only by way of attunement to chemical and physical law. As any competent chemist will tell you, the laws of chemistry and the chemical processes that are governed by these laws are patently not arbitrary or incidental. Indeed, the illustrious science of chemistry and all its manifold practical applications exist only because of the specific sensibleness of chemical processes. Thus, a wizard chemist able to whisk up various proteins and enzymes to order is no different in kind to the metabolic machinery of a cell, which can achieve the same feats. In both cases, the wizardry is possible only because of the sensible and intelligible laws of chemistry.

The rather striking conclusion reached is that the evolution of biological innovation is caused not simply by chance mutations, as Monod suggests, but by the contextual influence of Nature, in particular the laws of Nature and the various sensible and provocative effects of those laws upon matter and energy. Evolution is thus to be understood first and foremost as a *natural learning process* involving an accretion of sensible changes to genetic information over time. It is the naturally intelligent configuration of the system we call Nature that, through feedback, quite literally informs bio-logic. In this way, life learns how to behave in clever ways. Nature must be seen as a naturally intelligent system on every level, from its lawful and sensible configuration to the naturally intelligent tree of life spawned by this configuration. Holding these preliminary ideas in mind, let us continue.

CHAPTER 5

OSCILLATING PARADIGMS

At this point in our bold venture a tight and formal definition of natural intelligence as it pertains to life and evolution can be put as follows: *Evolution is a naturally intelligent learning process that weaves together naturally intelligent systems of bio-logic wise to their surroundings. This natural intelligence can be considered unconscious or even subconscious.*

In comparison with orthodoxy, this reasoning may sound rather fanciful. And yet however we choose to describe it, the evolution of life has clearly led to remarkably intelligent behavior. Indeed, how else are we to appraise the behavior of bio-logic if not in terms of "clever" or "smart"? If the reader recalls what I said in the prologue about hemoglobin being a tool with which to grab and transport oxygen molecules around the body, then consider also that hemoglobin needs to give up its cargo of oxygen when it has arrived at a specific location in the body. To do this, for hard-won oxygen to be suddenly liberated from its binding embrace with the iron within hemoglobin, yet more wizardry is required. It is all extremely impressive and clever. Even a casual Google search will verify that this single and seemingly trivial example of bio-logic is crammed with a kind of slick skill difficult to comprehend in full. If it takes a bit of a genius to fully comprehend the total functional repertoire of hemoglobin, this is only because the bio-logic in question is itself ingenious. In other words, life is really a *supra-intelligent* phenomenon. To be alive, to be able to stave off death and decay, to be able to make sense, *to be,* reflects the operation of natural intelligence. All of us live and breathe according to the intelligence embodied within the human organism, each component structure of which owes its existence to evolution, a process that *itself* owes its existence to the self-same principle of intelligence. Natural intelligence is everywhere: moving, flowing, and reverberating in ways sublime and spectacular.

Having said that, most evolutionary biologists aggressively contrive to negate natural intelligence. No matter what science comes across in its study of the natural world, no matter what science details in the esteemed pages of the journal *Nature,* its namesake, the very source of scientific data, is always denied intelligence. Not only is the process of evolution denied intelligent characteristics but organisms—the products of evolution—are likewise not generally spoken of as embodiments of intelligence. Compare this with the latest AI machines and computer hardware/software, which *are* talked of in terms of being smart and intelligent. These days, they even have intelligent shoes—so-called because an inbuilt computer chip can adapt the sole to different terrains. Very clever. And yet organisms that can adapt their shape according to differing environmental conditions (think of a vine trailing around an iron railing or the lens of an eye changing shape so as to achieve focus) are flatly denied any degree of intelligence. Indeed, the persistent and uncompromising efforts of the science community to avoid natural intelligence and the implications thereof are often unintentionally embarrassing. Just as people are prone to awkwardly shuffle their feet when it comes to taking on board some patent truth, some scientists will likewise awkwardly shuffle their concepts and their language in order to overlook the intelligence of Nature. Which brings us back to the modern art of merelyism. Here is Richard Dawkins using a rather unseemly linguistic technique with which to deny natural intelligence. In order to forcibly promote his merelyism, Dawkins has invented the word "designoid":

Designoid objects are living bodies and their products. Designoid objects *look* designed, so much so that some people—probably, alas, most people—think that they *are* designed. These people are wrong.[1]

Not so. People intuitively know that biological structures are designed in some way. Not by us, of course, but by Nature. In terms of structure, function, and behavior, organisms encapsulate a nifty form of design that is the product of millions upon millions of years of gradual but inevitable evolutionary honing. Nature, understood as an intelligently configured system, serves as the contextual impetus that evokes and drives the design. But Dawkins will have none of this. His determined negation of natural intelligence continues:

At the end of many generations of cumulative finding [evolution through natural selection], a designoid object is produced which may make us gasp with admiration at the perfection of its apparent design. But it is not real design, because it has been arrived at by a completely different process.[2]

Can you imagine the scene? We take a child by the hand and lead him or her through an autumnal forest. Stopping beneath the shade of a large leafy sweet chestnut tree, we pick up one of many recently fallen green seed capsules. The child marvels over the spiky seed case and is momentarily shocked when the spikes prick our fingers as we begin prying the seed case open. We explain that this spiky structure of the chestnut pod ably protects it from the unwelcome attention of potential consumers (we might also explain to the child that the cut engendered by the spiky seed case will heal soon enough since the substance fibrinogen in blood reacts with air to form a clot, thus stopping further bleeding).

Wounds aside, upon prizing the seedpod open, we point out that the several polished chestnuts nestled inside contain all the instructions necessary to grow another chestnut tree, including instructions on how to unzip DNA, replicate DNA, error-check DNA, sequence amino acids, build enzymes, manipulate proteins into mechanical micromachine architecture, sense gravity, locate light, covert light into food, sequester water and minerals, respond to seasons, respond to disease, and even produce copious amounts of similar seeds. Then, to cap it all, in true Dawkins style, we explain to the child that all these eminently sensible properties of the organic material are merely "designoid," *as if intelligence had nothing whatsoever to do with the manner in which huge amounts of specific information are built up into organisms.* We might warn the child not to gasp *too* much in wonder of organic life . . .

This is the implicit aim of the term "designoid" introduced by Dawkins. It is a rhetorical technique employed so as to stop us from acknowledging natural intelligence. Dawkins is warning us not to get too involved, not to let our emotions get the better of us, almost as if Nature and evolution were a bit *sneaky,* a bit *too* cunning, and well versed in the art of deceiving us humans. Designoid. Mimicking design. Not real design.

So what is real design? Human design alone? Done always with conscious aforethought and conscious intention? Never by way of trial and error of some unconscious kind? Never by way of context and fit? If we resolutely insist that design is a manifestation of human intelligence alone, then we are adopting a rather arrogant anthropocentric stance (as we do in our treatment of other faculties like learning, invention, innovation, and technology). But why, exactly, should our brains be endowed with information-processing capacities alien to the rest of Nature? Why should mental functions, based as they are in the brain, be utterly unrelated to the way the rest of Nature works? If human intelligence is all about learning and making sense through accessing and manipulating information, then evolving life likewise involves the operation of intelligence in terms of the constructive and sensible manner in which information about the environment is accessed and put to good use. All you need to do to differentiate between our information-processing abilities and Nature's information-processing abilities is to place the word "natural" before intelligence. So too with the notion of design. The chief difference between natural design and human design is that we plan ahead whereas Nature always operates in the living moment, preserving that which makes sense in the here and now (but even this moment-by-moment design can nonetheless facilitate a kind of anticipation—witness the aforementioned fibrinogen, for example, which is primed to stop bleeding, even if wounds may never arise).

Similarly, why monopolize words like "invention" when something like, say, flight was obviously invented by life through the evolution of wings? Rather than coining more horrible words like "inventionoid" when accounting for life's ability to be inventive, why not simply speak about natural invention? After all, Darwin regularly had recourse to this sort of terminology. Indeed, no one seems to have a problem with Darwin's term "natural selection"; no one is calling for the term "selectionoid" to be applied. Thus, why Dawkins could not simply talk about natural design instead of coining the ugly term "designoid" is rather baffling. Natural design/intelligence and human design/intelligence are not mutually exclusive phenomena—each depends in some essential way upon the non-arbitrary organization of information. Human design can be seen as basically an extension, or *conscious refinement,* of previous means of design.

Even if we bang our fists on the table and assert that design absolutely must be synonymous with intention, then Nature, as a whole, can be viewed as a kind of intentional system, just as a genetic algorithm substantiated within a computer system so as to evolve, say, a face-recognition program can be viewed as an intentional system. In each case, the intention, the will to design as it were, is inherent in the lawful context of the system. Lawful contexts ensure that specific (evolutionary) events will happen within them. This does strongly imply that Nature is associated in some way with a "natural intention" (unsettling as that may sound to some), evolution representing the inevitable method through which such a "natural intention" is realized. As we shall see in more detail later, this "natural intention" would appear to be bound up with sense-making, whereby Nature makes more and more sense of itself, or reflects itself more and more accurately, through the evolutionary process.

Once more, then, it seems that even if we ignore for the time being the awesome natural design of all life-forms, the existence of natural intelligence as a design force becomes even more apparent when taking into consideration the larger context according to which evolution is provoked and sustained. However, Dawkins reveals the fact that, like others in his field, he has paid scant attention to those contextual factors that elicit and encourage evolution. This is like conveniently ignoring the context of human intelligence that, at a fundamental level, governs computer simulations of evolution. In the following statement, Dawkins echoes the sentiments of Steven Levy detailed in the last chapter:

> Natural selection is an extremely simple process, in the sense that very little machinery needs to be set up in order for it to work. Of course, the effects and consequences of natural selection are complex in the extreme. But in order to set natural selection going on a real planet, all that is required is the existence of inherited information.[3]

And here's the rub. If you can spot Dawkins's mental shortfall, you can once more begin to appreciate natural intelligence. Natural selection only *seems* extremely simple. In reality, a system has to be specifically configured in order to elicit natural selection. This specific configuration—the necessary "machinery" as Dawkins calls it—is far from simple, just as the machinery needed to evolve the virtual fish eye was far from simple. After all, modern computing technology along with a fair whack of programming ingenuity was required to elicit the virtual fish eye. Once the appropriate system is set up, only then will evolutionary miracles be wrought, only then will things appear simple and undesigned, only then can we sit back and watch creative magic unfold before our eyes.

More to the point, where do planets and vehicles of inherited information come from? You need law-conforming matter (evinced in the sensible structure of the periodic table) and a precise law of gravity to name but two obvious ingredients. And lots of constructive carbon to boot (not to mention water and the untold amazing properties of water). As any physicist will tell you, matter and gravity are hardly devoid of sense and lawful order. They are not simple circumstances to be casually brushed aside as if they were of no contextual import in accounting for planetary life. Granted, once specific laws are in place, in an axiomatic sense, then further things can happen, but such laws must be *given,* a sort of highly specified set of ground rules according to which creativity can emerge. And it is not even enough to have some primitive self-replicating molecule (like RNA). For evolution to really take hold that replicator has to have a significant *encodable* relationship with building blocks (like amino acids and proteins). In other words, you need an elaborate and astonishingly flexible Lego-like system to work with. All this is very far from simple.

Moreover, any given planet is still not sufficient to foster evolution. You need a long-term source of energy, namely a sun. And are we to dismiss a sun or star as merely nothing much? Simply a trivial bright round object up in the sky that anyone could cobble together in his or her spare time? Careful observation reveals that stable and enduring suns with their unique life spans (the supernova part of which generates carbon and all the other heavier elements so crucial for life) exist and persist only because of the precise laws of physics (like gravity and the nuclear forces that bind atoms) and the precise behavior of hydrogen, both quantum and atomic. The stunning result of these natural laws in the case of our sun is that it radiates high-grade energy equivalent to a million ten-megaton hydrogen bombs each and every second. This is patently not a minor phenomenon to be taken for granted, but part and parcel of the exquisite fabric of Nature. In fact, in tracing the existence of planets and stars back to the hypothetical big bang, we find that no one thing can be fully explained without evoking other things. Absolutely nothing is isolated. All things, in terms of their meaning and how they behave and react, depend upon all other things. A contextual web of law-conforming relations comes to mind once more. So, in order for Dawkins's eventual simple system of "inherited information" to come about and launch itself into the spectacular tree of life, a whole host of *complex* and *specific* interconnected events and sensible contextual laws are required. As Carl Sagan wryly noted, in order to make a "simple" apple pie, first you must make a Universe. So Nature, and the non-random lawful context of Nature, can clearly not be casually brushed aside as an incidental part of evolution. One ignores contextual forces such as laws at one's peril (perhaps, if pushed, Dawkins might concede that the Universe itself is "designoid"—but this is more an issue of words and definitions than of objective facts).

TREES OF LIFE

Although it is most evident in the evolving tree of life, I have repeatedly alluded that intelligence is also apparent in the laws of Nature—that the contextual influence of the laws of Nature represents a kind of "primary expression" of intelligence. This can be further appreciated if we think about the growth and development of a real tree. A tree's structure is such that it branches out in all directions and produces leaves that orient themselves toward ambient sunlight. The shape and poise of leaves pays testimony to this sunlight immersion, with all the leaves of the tree adding up to a distributed surface able to "experience" as much sunlight as possible. This makes good sense, of course, and plants excel in this kind of botanical intelligence. Now, it is clear that what is driving the evolution of specific leaf structures and specific leaf distribution is sunlight itself and the energetic potential of sunlight. Which is to say that ambient high-grade solar energy can be made use of, can be made sense of, can be fed upon, can be harnessed, so long as the right method is utilized. If the energetic photons that make up sunlight can be "captured" and be put to work in molecular manipulation, then metabolism can be achieved.

In other words, leaves and the nanotechnological photosynthetic mechanisms inside them are essentially making sense of the sensible properties and sensible order embodied within sunlight (just like the pointed shape of leaves makes sense of the orderly flow of water). Or, to put this in another way, trees and their leaves represent a form of biological intelligence responding to the law and order of sunlight. The thing is, while we can, if we so wish, readily appreciate the organic intelligence within leaves and their busy photosynthesizing components, it is somewhat harder to grasp the fact that this kind of biological intelligence exists only because of the intelligible context afforded by light (and other aspects of the environment). Trees and plants, or at least specific parts of them, are quite literally built around, and upon, the sensible order inherent within light energy. One kind of order feeds upon another kind of order. One kind of sensible behavior feeds upon another kind of sensible behavior. Moreover, considering that sunlight depends upon a stable star and that stable stars depend upon stable laws, it seems logical to infer that everything is tied together into one eminently sensible system. In the last analysis, one needs a lawful and sensible Universe to generate suns, planets, and evolving life. Leaves thus feed off sunlight, which itself feeds off natural law, natural intelligence being bound up with all three phenomena. In summation, one can say that the self-organizing intelligence embodied in living things must be matched by the self-organizing intelligence existing all around them.

If the reader still thinks this reasoning to be fanciful, then bear in mind that with artificial life and artificial evolution we have firmly established that digital organisms are, in the ultimate analysis, the embodied reflections of the human intelligence that, willy-nilly, set their virtual worlds into lawful software-driven motion. Seen as a whole, a computer system running a simulation of evolution in which smart digital entities emerge has intelligence written all over it, in every wire, chip, and piece of machine code. Such a principle in which intelligence begets intelligence carries over into the real world.

DEALING WITH DANIEL DENNETT

It is with some degree of consternation that one learns that Dawkins's forceful and influential sympathies are matched by any number of other eminent thinkers. Take Daniel Dennett, for instance, an esteemed philosopher who helps promote Dawkins's merelyistic perspective on life. Author of the hefty and oft-cited five-hundred-page treatise *Darwin's Dangerous Idea,* Dennett explores evolution in all its twists and turns and attempts to scupper any vestige of intelligence within Nature. In fact, Dennett attempts to use the theory of evolution to account for everything of interest in the Universe. However, although convincing in much of his rhetoric, Dennett suffers the same conceptual myopia as Dawkins, namely the inability to take into account the context-dependent aspects of evolution. Well, to be fair, Dennett does realize that context is important in understanding the evolution of life, yet, as we shall see, he still manages to craftily dodge any invocation of natural intelligence. Then again, the kind of short-sightedness evinced by both Dawkins and Dennett serves to make natural intelligence come even more sharply into focus, so their views are actually quite useful to our ends.

Dennett embarks upon his merelyistic crusade by defining evolution as an algorithm. Recall that an algorithm is a computational procedure that works out, or produces, something specific. That organic evolution is algorithmic is implied, as we have seen, in the science of genetic algorithms whereby a basic evolutionary process can be manifested within a computer system. Left to evolve for long enough, logic dictates that a population of digital entities subject to generate/test software will eventually evolve to make maximum sense of their virtual environment. However, while this algorithmic view of evolution might be sound, the tendency is to ignore the various contextual factors that govern algorithms and endow them with design potential. Here's Dennett showing off his blind eye:

No matter how impressive the products of an algorithm, the underlying process always consists of nothing but a set of individually mindless steps succeeding each other without the help of any intelligent supervision.[4]

There it is yet again, the monotonous "nothing but" approach to evolution. And mindless of course. But it is an illusion to think of an algorithm as mindless since, as we have seen repeatedly, to actually incarnate one you need to intelligently supervise its configuration. Once a smart configuration has been arranged by intelligence, a measure of intelligence is then inherent in the system and will gradually become reflected in some new form via an evolutionary response to it. Well-crafted order of one kind is needed to craft order of another kind. Intelligence begets intelligibility begets intelligence.

In the real world, Nature itself, as a whole system, supervises and orders the evolution of bio-logic. Although this self-organizing process invariably produces things that are redolent in natural design—like us, for example—for Dennett these are just fortuitous side effects of the dumb algorithm that is evolution:

The theory of natural selection shows how every feature of the natural world *can* be the product of a blind, unforesightful, nonteleological, ultimately mechanical process of differential reproduction over long periods of time.[5]

Well yes, but only if we choose to conveniently ignore the contextual factors that allow evolution to flourish. By ignoring the greater context within which evolution is born and proceeds, we are ignoring the very causal factors involved in the thing we are trying to explain—namely the design finesse of living things. To repeat, one cannot fully explain evolution without recourse to explaining why and how evolution comes to be. To say that it "just happens," or that it is "just an algorithm," gets us nowhere and stops us looking at the bigger picture. To ignore the origins of the greatest and most fantastic process we know of—ignoring evolution's contextual backdrop, as it were—is akin to marveling over the digitally evolved creations running amok inside a virtual reality and then ignoring how the software and hardware running the show came to be in the first place.

It should also be pointed out that, like most evolutionary thinkers, Dennett reveals in the above quote that he is blind to the inherent directionality of evolutionary processes. Remember, evolution is always about the making of sense, wherein selective feedback drives a closer fit of a replicating object to its context. As an organism or a digital object evolves, it will invariably come to make more and more sense of the environment in which it is embedded. It *has* to do this because only sensible context-dependent behavior can confer fitness. Through repeated selection, an environment will "drive sense" into replicating media. That which makes sense will be more likely to be selected. That which makes sense will be more likely to survive. Conversely, that which fails to make sense will be more likely to die. Hence there is a target within evolutionary processes. The target is inherent within the system.

THE GAME OF LIFE AND BEYOND

Dennett goes on to consider the Game of Life, a popular computer algorithm invented by mathematician John Conway in the early 1970s. In the Game of Life, lifelike digital entities are yielded, which roam around the computer screen, and some of them can reproduce themselves. The remarkable thing about the Game of Life is that the digital entities emerge in response to but a few "simple" local rules, which are reiterated again and again in true algorithmic style. Dennett explains that Conway chose these rules out of a possible infinity of rules. Of all possible rules, the few chosen by Conway are those that work and lead to interesting events that can entertain us as they unfold.

In comparing the real system of life with Conway's Game of Life, Dennett first suggests that any proposed "divine" intelligence present within the Universe need only be a *lawmaker*. In other words, such a divine intelligence need not create smartly designed creatures out of thin air but need only find the *right laws* to allow for an evolutionary algorithm to flourish (thus allowing natural selection to do all the design work), just as Conway alighted upon the right laws to induce the Game of Life into action. The "right laws" are, of course, synonymous with the "right context." Once the right lawful context is established, evolution can arise and then design complex things.

In an apparent coup de grâce, Dennett next argues that since anyone could conceivably have alighted upon the significant rules governing the Game of Life (through trial and error), the divine lawmaker becomes promptly demoted to a divine *law-finder*. You can see what business Dennett is about. Although perhaps smelling a bittersweet (to him) hint of natural intelligence, Dennett is quick to reduce any such nonhuman intelligence further and further away from the theoretical picture. But how on earth to be rid of a pesky divine law-finder, or, in our parlance, how to be rid of an intelligently configured context (i.e., an intelligent Nature) according to which life can evolve?

In this most important of conceptual problems, the nitty-gritty as it were, Dennett suffers an attack of what can only be described as a kind of merelyistic madness. Whilst his attempts at dismissing natural intelligence can conceivably be considered somewhat laudable in terms of fulfilling the egotistic desire to keep human minds as the sole possessors of elaborate intelligence, Dennett's final dodge is nonetheless absurd in the strongest sense possible, a worse dodge even than Dawkins's ad hoc creation of the word "designoid." Here is how Dennett finally comes to terms with things like the fine-tuned laws of physics and chemistry, which, for us, are part and parcel of natural intelligence:

> What would the Darwinian alternative . . . be? That there has been an evolution of worlds (in the sense of whole universes), and the world we find ourselves in is simply one among countless others that have existed through eternity.[6]

Say what? Are we to explain away the finely tuned context of Nature by appealing to the so-called multi-Universe theory? (Even Dawkins has embraced such an ultra-convoluted scenario in his book *The God Delusion*.) Note that the multi-Universe theory holds, in no uncertain terms, that an infinity of (unobservable) Universes exist, perhaps birthing one another in some evolutionary manner. Not a million other Universes, mind you, not even a billion trillion zillion squillion, but an infinity. That's more than a lot (more, in fact, than is truly conceivable). If so, then *all* contexts exist and we just happen to be in one of those infinity of Universes where the context *appears* intelligently configured so as to foster evolution to the point of self-reflective consciousness. Indeed, there will also be an infinity of Universes just like ours (with a significant lawful context ensuring life) but with minor differences such as how thick Dennett's beard usually is. Taking this infinity principle to its most (but still theoretically viable) extreme, there will even be more than ten billion Universes in which some immutable principle of physical law causes Dennett to "pop" miraculously out of existence at some definite time, perhaps when you finish reading this sentence. Anything and everything goes in the multi-Universe paradigm. There can be no reins on our imagination whatsoever, since every possible physical law is realized. Unless, of course, there is some mysterious "higher law" that stops certain peculiar and untoward laws from manifesting. But then this "higher law" would itself had to have arisen from an infinity of possible higher laws. And so on all the way to the loony bin.

What an inelegant dodge is this multi-Universe scenario. Is the threat of natural intelligence really that great? Is the need for the human ego to monopolize intelligence, design, and purpose really that crucial? Apparently so. Indeed, not only does Dennett introduce the above statement sharpishly and without much warning, notice that he also slips in an appeal to eternity. As if he was not invoking much! Yet his words (which include "simply," of course) come across like some casual aside (I am surprised that he did not just append the loaded statement as a brief footnote!). While it is conceded that an appeal to eternity is, as we shall appreciate in the last chapter, somewhat inescapable when dealing with the ultimate "why" of the Universe, to invoke it in such cursory fashion without further elaboration as part of a multi-Universe theory is to use the concept as an *excuse,* or escape button, as opposed to an explanational device with which to aid our understanding.

There are any number of reasons why an appeal to a multi-Universe scenario is nonsensical. Not only are we passing the explanational buck to where it is ostensibly, but lumpishly, out of sight, such a conceptual move also patently ignores the use of Occam's Razor. The formidable cutting edge of Occam's Razor warns us that we should always stick to the most parsimonious explanation for any given phenomenon. To imply that an infinite number of Universes running for eternity and for no reason be the ultimate explanation for the existence of life represents no less than the most blatant disregard for Occam's Razor conceivable. Occam would not just turn over in his grave upon hearing of this multi-Universe notion, he would verily spin every which way.

In fact, if one appeals to the multi-Universe scenario to explain the fine-tuning of Nature, then one need never explain anything again. Science and intellectual inquiry become redundant. Whatever we come across, whether in the physics lab or within some ecosystem, we need only shrug and state nonchalantly that we just happen to live in that particular Universe among countless others that just happens to have produced that particular phenomenon. It is an all-and-everything-goes approach to life. All and any significance to existence can be casually brushed aside as simple flotsam and jetsam floating in an infinite sea of Universes in which all and every eventuality is manifest somewhere and somewhen. And if, God forbid, one dares to ask a proponent of the multi-Universe theory *why* an infinite number of Universes should exist, no answer can be given except "why not?" This is not an answer but only more proof, if more proof were needed, that the multi-Universe approach to understanding our existence is little more than devious buck-pushing. Nothing is explained. All that has happened is that an infinite amount of more stuff is now in need of an explanation. Worse still, the multi-Universe theory represents a kind of modern-day insubstantial religious dogma, for it is evident that one must have faith in that which can never really be verified. I submit that such an infinitely bloated paradigm is nothing more than a conniving means by which the fragile human ego avoids having to acknowledge natural intelligence. (And let me here preempt any accusation that the natural intelligence paradigm entails the need to explain something over and above the Universe. Since natural intelligence essentially *is* the Universe, an attempt to explain the origin of natural intelligence is actually no different from conventional attempts to explain the Universe's origin.)

FACING UP TO NATURAL INTELLIGENCE

To meet this extraordinary claim of Dennett's, slipped into the text as if by sleight of hand, is surprising to say the least. Prior to this assertion, it is clear that Dennett has been skirting the issue of natural intelligence (in his own way, of course), and it is obvious that he is keenly aware of the significant context of Nature that facilitates evolution to the point of conscious brains (although I am not sure he has ever thought much about the genetic code, proteins, and the Lego-like nature of reality). But it is still shocking to see one so hallowed in the domain of philosophical inquiry resorting to the multi-Universe theory, a theory that attempts to explain something by multiplying that something to an infinite extent. Really, there is no end to the multi-Universe theory. An infinity of all possible Universes might be further subsumed under an infinity of all possible infinities. And so on ad nauseam.

Clearly, those who propound some kind of multi-Universe theory—and there are any number of convoluted variations—feel themselves to be floundering in expletively deep metaphysical water, their situation perceived to be so dire as to force them into invoking more than is humanly imaginable. And it must be said that the root cause of such unbounded speculation appears to lie with the overwhelming desire to banish all and any indications that Nature be teleological (i.e., purposeful) and, by implication, intelligent in some way. This approach leaves the human ego intact and leaves human intelligence and the human mind as the only really refined design agency in Nature. (I am reminded here of theories that state that the fine-tuning of the Universe indicates that we must be living inside a vast computer simulation designed by advanced beings in some real Universe. Obviously this theory has not been thought through. For the creators of the simulation would themselves have to be living in a similarly fine-tuned Universe and would therefore have to conclude that they too were living in a simulation—thus leading to a classic infinite regress.)

While it can be granted that the paradigm of natural intelligence is equally as astonishing as the multi-Universe paradigm, it surely makes more sense than the multi-Universe theory. At least the paradigm of natural intelligence—which states that intelligence is a fundamental property of Nature—is more compact and deals exclusively with the only Universe with which we are familiar and that we know to be real. To attempt to account for the existence of something (our sensible and evolution-friendly Universe) by invoking an infinite amount of unobservable somethings (an infinity of other equally real but unobservable Universes), which are deemed to exist for no reason whatsoever, is to flee from the something in question, the very thing we wish to explain. And why should one flee in such a dramatic style, with such an infinite amount of baggage tucked under one's shaky arm? Only because the consequences of the truth—that Nature is imbued with intelligent characteristics—are simply too hard or too humbling to swallow. And even if one does entertain the multi-Universe theory, in the last analysis *nothing really changes,* for if an infinity of Universes really does exist, then *at least one will exist in which natural intelligence governs everything.* With an infinity of Universes there will be one (well, actually an infinity!) that comes complete with intelligence encoded into its lawful fabric. It would be "intentional" and intelligently behaving in some way right from the start.

MOOD SWINGS

I'll wager that there is something about this flagrant negation of natural intelligence that relates to modern culture, that we are living in an age in which there is a mighty backlash against religion and the sometimes trite beliefs associated with it. Having become more intellectually sophisticated through science in the last few centuries, and having grasped how much of Nature works, we modern humans have convinced ourselves that we can explain everything away. All and any indication that Nature bears intelligent characteristics—and anyone who takes the time to look deeply at evolution is perforce compelled to sense intelligence, whether in organisms or in the laws that allow for the evolution of organisms—is curtly dismissed, presumably because it is considered to have the same flavor as religious beliefs.

While much of the drive away from religious dogma is justifiable on the grounds that science works remarkably well without overtly depending upon unchallenged faith, and while science has provided us with a much greater understanding of Nature than most religious ideologies, this has not lessened the wonder of the Universe in any way. Indeed, the wonder of Nature as revealed by science is something to be warmly celebrated. As Carl Sagan was fond of pointing out, the Universe as revealed by modern science is actually *more* grand and *more* fantastic than the majority of religious beliefs about the Universe.

Even so, natural intelligence is still *not* acknowledged. Moreover, newly discovered aspects of natural intelligence are received by the scientific community in such a way that it appears as if the scientists who have made the discovery actually *invented* the natural intelligence themselves. For example, Watson and Crick are still revered to this day for their pioneering studies of DNA in the 1950s and their subsequent elucidation of the genetic code. But surely Nobel prizes and academic acclaim are dished out to the *wrong* party. For it is Nature and the constructive intelligence of Nature that determines all the fantastic things that science discovers through its ongoing dialogue with Nature.

Watson and Crick did not invent the exquisite and elegant double-helix structure of DNA, and they did not pen the genetic code; rather they merely managed to *uncover* it and then observe its ingenious embodiment of, and reflection of, natural intelligence. The same holds true for the Wright brothers, who have become legendary icons, whereas the bird wings they based their crucial aerofoils on remain uncelebrated.

The human race excels in its refusal to acknowledge natural intelligence. Consider the genetic code. Have a real good think about it. A *code*? Nature shaped a code? How smart is that? Yet, recoiling from thousands of years of unquestioned religious indoctrination (with all its subsequent wars and misunderstandings), we resolutely refuse to admit natural intelligence into our conceptual paradigms about the world. Since an intuition of natural intelligence might smack of religion, of some intelligence apart from our own, we may choose to label the idea as religious mumbo-jumbo and do everything within our power to banish the notion, to cover up all and any traces of the very fine intelligence embodied in both living things and the laws that provoked them to arise. It is surely high time that this matter be rectified, that we courageously adopt and explore the natural intelligence paradigm. Nature, the great system that engendered us but from which we have become so terribly alienated, is surely deserving of such a change in attitude.

THE IGNOBLE ART OF
DENYING NATURAL INTELLIGENCE

Take any textbook about some natural phenomenon, say a neuropsychology textbook or a textbook about embryology. In each case, the book will, if it is detailed, be hard to understand in full. One might consider the authors to be very intelligent. In their descriptions of the way the brain organizes information by way of firing neurons or the way in which a fetus develops organs and other physiological structures, the authors' grasp of such biological processes might leave us feeling stunned. Yet the text is wholly a reflection of natural intelligence, the attempt to translate the naturally intelligent behavior evinced by bio-logic into a worded form. In this sense, natural intelligence has actually taught science all that science knows. A physicist, a chemist, a biologist, a geneticist—all are informed by natural intelligence. Indeed, the entire enterprise of science and the rush for PhDs and scientific acclaim is built upon the knowledge and learning afforded by Nature, the knowledge distributed and manifestly realized throughout the natural world wherever we care to look and prod. More to the point, we are able to seek and grasp this knowledge only by virtue of the human nervous system, which itself embodies and testifies to the causal presence of natural intelligence.

Thus, if we are honest, we must face the humbling fact that the greatest thing on this planet is not human intelligence or the products of human intelligence, for both these phenomena exist only because natural intelligence exists. Without natural intelligence there would be no laws, no self-organizing matter, no DNA, no evolution of DNA, no speciating organisms, no biosphere, no human race, and no human intelligence. Natural intelligence, by definition, must *precede* human intelligence. It is the prime mover, the very cause of life's existence. Without its capacity to realize evolution and thence extend the thrust of evolution into the complexification of the hominid cortex, we would not be here in such extraordinary conscious circumstances.

EVOLUTION

The Most Vivid Expression of Natural Intelligence

Our man Darwin was worried about the theory of evolution offending the religious sensibilities of his peers, since God was no longer needed to miraculously create organisms out of thin air. In reality, Darwin was disseminating the logical principles by which Nature works its wonders. This was an intellectual advance beyond simple religious dogma, which saw the Creator whisking up the tree of life in a few days of hard labor. In any case, evolution is not something to be angry about or loathe (one gets the feeling that many religious people *hate* the idea of evolution).

In terms of the novel creativity and novel orchestration of organic molecules, Nature excels in its evolutionary endeavors, the human species being but one instance of this astonishing form of natural organismic creativity. Yet even though "God," as a supernatural omnipotent patriarchal personal deity, is now an outdated and redundant concept, and even though science now provides us with an even better picture of the Universe than our religious forebears could (i.e., a more detailed and more accurate picture), this new picture, despite obvious signs to the contrary, is nonetheless interpreted as being wholly devoid of intelligence and purpose. Which means that in superseding a dogmatic belief in a personal Almighty, we have "thrown the baby out with the bathwater" and become blind to natural intelligence, a very real aspect of Nature that science cannot help but document, especially in the study of evolution and the contextual factors necessary to induce evolution. So while we now understand much about DNA and the evolutionary modification of DNA, we fail to realize the implications of such an incredible state of natural affairs. Natural intelligence is looming before us if we but care to open our eyes and look beyond the reactionary merelyistic dogma espoused by evolutionary science in its response to centuries of religious thinking. To perceive natural intelligence is to move beyond both dogmatic religion and dogmatic science, to a vantage point where the intelligence of Nature becomes very much apparent. (And although I avidly avoid the term "God," a new pantheistic understanding of "God" is not out of the question.)

WHAT DOES DARWIN ACTUALLY HAVE TO SAY?

At this juncture, it would be quite impressive to quote Darwin himself on those aspects of Nature that suggest intelligence. Most people assume Darwin saw no design or plan in Nature whatsoever, as if he had reduced the web of life to a mindless automatic consequence of an equally mindless and automatic process. But Darwin, having outlined the theory of evolution so lucidly and so expertly in *The Origin of*

Species, must surely have dwelt long and hard on the implications of the evolutionary process, a process that he probably contemplated more than anybody else in history (it took him fully twenty years to proceed with publication). If Darwin had thought about it deeply enough, he would certainly have realized that evolution had to be "set up" by Nature in the first place. He would have been forced to consider the contextual status of Nature, the lawful aspect of which permits and provokes evolution. Actually, the word "Nature" is one of the most commonly used words in *The Origin of Species.* Natural selection, the survival of the fittest, the struggle for life, the emergence of new species—all these processes are subsumed under the ever-vigilant eye of Nature, the very contextual system that ensures both the emergence of heritable information (DNA) and the ongoing elaboration and complexification of that information.

In point of fact, we must assume that Darwin *did* come to the conclusion that Nature is intelligently configured. In a (little-known) letter to fellow naturalist Asa Gray, Darwin wrote: "I am inclined to look at everything as resulting from *designed laws,* with the details, whether good or bad, left to the working out of what we may call chance"[7] (my italics).

Had Darwin lived in our modern digital age, and had he been privy to the evolutionary potential of genetic algorithms, it would surely have occurred to him that any given genetic algorithm must be somewhat designed in advance (in both software and hardware). This realization would have provided Darwin with more support for his statement regarding designed laws. Of course, this does not necessarily mean that Nature was itself designed by some other intelligence, but rather that Nature is an intelligence unto itself, Nature's fine-tuned laws representing the most primary expression of this intelligence (as I intimated earlier, natural intelligence *is* the Universe).

Whilst we are considering this issue, I might as well quote another scientific genius who also felt that Nature in its entirety was infused with intelligence. Albert Einstein spoke of his great reverence for Nature, whose intelligibility and comprehensibility provoked in him an emotion that took the form of:

a rapturous amazement at the harmony of natural law, which reveals an intelligence of such superiority that, compared with it, all the systematic thinking and acting of human beings is an utterly insignificant reflection.[8]

Contrast this with the view of Dawkins (who seems always to play down Einstein's pantheistic sentiments) and Dennett, two intellectual giants of our day, both of whom appear to have used their allotted portion of human intelligence to smother any inkling that natural intelligence exists. This situation becomes decidedly ridiculous when we consider that the only reason they are able to wax so merelyistic in the first place is because they are both blessed with one of the most extraordinarily complex "devices" that we know of, namely the conscious human cortex. Natural intelligence has fostered this astounding evolutionary turn of events and yet they still manage to assert that their being, their condition of being alive and able to cogitate, is somehow the product of a non-intelligent process. Which means, in this case, that human intelligence is denying that it has intelligent roots. This is like building the most powerful computer system in the world, inoculating it with an ultra-smart genetic algorithm designed to eventually evolve a conscious artificial intelligence, and then finding that the artificial intelligence so evolved denies that it has emerged by way of any kind of intelligence or purpose. I think we would all be somewhat irked by such an artificial intelligence. And yet this is exactly the sort of hubris Dawkins and Dennett convey. Underscored by the hundred-billion-celled human cortex that is far and away more complex than any man-made computer system and miraculously conducive to conscious processes, the kind of conscious intelligence afforded to both Dawkins and Dennett nevertheless concludes that it has arisen due to a non-intelligent process. Let's face it; our take on evolution is in dire need of readjustment.

While we are on this subject, further evidence of the ultra-smartness of human biology (such as is evinced in the human cortex) can be cited. The reader is here invited to be impressed by the following statement by T. S. Ray, the artificial life researcher we met in the last chapter (computer buffs will likely be more amazed). While Ray's description of the inner workings of the human organism is overtly computational, it still serves to highlight the *extreme* intelligence evinced by bio-logic:

It is generally recognized that evolution is the only process with a proven ability to generate intelligence. It is less well recognized that evolution also has a proven ability to generate parallel software of great complexity. In making life a metaphor for computation, we will think of the genome, the DNA, as the program, and we will think of each cell in the organism as a processor (CPU). A large, multicellular organism like a human contains trillions of cells/processors. The genetic program contains billions of nucleotides/instructions.

In a multicelled organism, cells are differentiated into many cell types such as brain cells, muscle cells, liver cells, kidney cells, etc. The cell types just named are actually general classes of cell types within which there are many subtypes. However, when we specify the ultimate indivisible types, what characterizes a type is the set of genes it expresses. Different cell types express different combinations of genes. In a large organism, there will be a very large number of cells of most types. All cells of the same type express the same genes.

The cells of a single-cell type can be thought of as exhibiting parallelism of the SIMD kind [single instruction multiple data—all CPU processors do the same things upon their data, even if the data is different for each], because they are all running the same "program" by expressing the same genes. Cells of different cell types exhibit MIMD parallelism [multiple instruction multiple data—CPU processors can be executing different code but all are orchestrated toward a common goal] as they run different codes by expressing different genes. Thus, large multicellular organisms display parallelism on an astronomical scale, combining both SIMD and MIMD parallelism into a beautifully integrated whole. From these considerations, it is evident that evolution has a proven ability to generate massively parallel software embedded in wetware.[9]

Quite. More reason then to start acknowledging the organic intelligence that saturates our being. This intelligence might move at a different tempo to human intelligence, it might be autonomous and unconscious, it might even manifest in distinctly alienesque ways, yet it is undoubtedly an "other" intelligence apart from human intelligence. And its presence is everywhere, not simply in organisms but in the very laws of Nature that encourage parallel processing systems of bio-logic to evolve in the first place.

THE LAWS OF NATURE

In speaking of natural intelligence in its fullest sense, laws invariably need to be invoked somewhere along the line. This explains my repeated references to the laws of Nature. Such an invocation is especially necessary when tracing the ultimate source of natural intelligence. By source, I mean those aspects of natural intelligence that, although somewhat intangible, nevertheless lead to all the interesting tangible things found in Nature. Definite and non-arbitrary natural laws are clearly needed to encourage and foster naturally intelligent processes. Indeed, natural laws are perhaps the most informative feature of the contextual system that is Nature. Without laws, reality would quite literally fall apart. What this means is that in any critical analysis of evolution, we will *always* be forced into a consideration of natural laws.

Let us quickly retrace our steps and remind ourselves why this is so. Organisms are tangible and observable embodiments of intelligence. We, for example, are most definitely built out of smart systems of bio-logic (or sets of interacting biological tools, to recall the prologue). If we were lucky enough to observe our cells under a good modern microscope, the elegant and purposeful behavior that they embody would become overwhelmingly apparent. And if we were as astute as T. S.

Ray we might even discern the parallel information processing carried out by large arrays of such cells. At any rate, the intricate metabolic functions being performed by cells in which millions of molecules are carefully manipulated and juggled so as to sustain the cell's autopoietic nature obviously indicate an ultra-smart technique in action (learned, of course, through evolution). Likewise with the organs and the entire organism of which those cells are a part. It is silly to mince words at this stage—it is all natural intelligence in fluidic action.

Now, if we ask ourselves why natural intelligence is in fluidic action, or why it is that the evolution of life is assured, we are forced to invoke context, for as we have repeatedly seen, evolutionary processes always transpire in a system that is contextually configured in a sensible and intelligent manner (configured, for example, such that sensible genetic changes are continually sustained and built upon). And it is here that we must confront natural laws. The tree of life is a startling effect of specific natural laws, just as our beloved virtual fish eye was a startling effect of specific software laws. The inevitable conclusion is that the ultimate source of intelligence must lie with the highly specific laws of Nature. The intelligence (of some kind) inherent in the specificity of these laws then reflects itself in the process of evolution and the biological products of evolution.

It might be argued here that all this talk of natural laws is deceptive, that natural laws do not in fact have any objective existence. Perhaps they are merely the inventions of science, residing solely in the minds of scientists, or even conceptual conveniences, allowing them to be written as a concise set of equations on a blackboard.

That the laws of Nature genuinely *are* objectively real aspects of reality and not simply man-made constructs has been much discussed by physicist and popular science writer Paul Davies. His conclusions seem reasonable enough:

> It is important to understand that the regularities of nature are real.
> Sometimes it is argued that laws of nature, which are attempts to

capture these regularities systematically, are imposed on the world by our minds in order to make sense of it. It is certainly true that the human mind does have an extraordinary tendency to spot patterns, and even to imagine them where none exist. Our ancestors saw animals and gods amid the stars, and invented the constellations. And we have all looked for faces in clouds and rocks and flames. Nevertheless, I believe any suggestion that the laws of nature are similar projections of the human mind is absurd. The existence of regularities in nature is an objective mathematical fact. On the other hand, the statements called laws that are found in textbooks clearly *are* human inventions, but inventions designed to reflect, albeit imperfectly, actually existing properties of nature. Without this assumption that the regularities are real, science is reduced to a meaningless charade.[10]

Davies goes on to describe four of the main characteristics of natural laws. First, these laws are universal in that they are deemed to operate everywhere and everywhen. Second, they are absolute (i.e., they do not depend upon anything else, such as the state of the system through which they yield their effects). Third, they are timeless or eternal. And fourth, they are omnipotent. If we accept this widely accepted interpretation of the status of natural law and add natural intelligence to the mix, then the laws of Nature are equivalent to natural intelligence in its primary and, as far as we know, most fundamental expression. Moreover, Nature's laws must surely represent the ultimate contextual influence since they effectively pervade and influence everything.

A fifth characteristic of natural laws can be invoked, namely their essential invisibility. This explains why we usually find it so hard to realize their objective presence. We are so used to natural laws, particularly their manifest effects, that we fail to really appreciate them. It's a bit like being impressed by neat patterns of iron filings on a sheet of paper without bothering to reflect upon the magnet underneath that caused the patterns to form. Were it not for the magnet there would be no patterns.

In the same way, we can be pretty much oblivious to the laws of Nature. Solar eclipses are a case in point. Most people are more impressed with an actual solar eclipse itself than an eclipse's timing, which will be accurately predicted (almost to the second) by science in light of science's knowledge of gravity and celestial motion. As a fundamental law governing all celestial motion and granting highly accurate predictions for future celestial events, the unwavering nature of gravitation should greatly impress us, yet the truth is that, as with others of Nature's great laws, we are generally oblivious to it (the same can be said of the laws that hold atoms in place). If a new natural law came mysteriously into force such that the sky turned from blue to pink, we would, of course, become very much aware of the new law. But the honeymoon of astonishment would be short-lived. Soon enough we would mentally acclimatize ourselves to the effects of the new law and would no longer be sensitive to the law's omnipresence and objective status. The same is true of software laws and their role in governing computations. While their effects may be virtually tangible in terms of an impressive computer display in front of us, the software laws underlying the display are essentially invisible. In whatever medium laws are at work, we are generally oblivious to their objective (albeit immaterial) nature.

Returning to the notion that the laws of Nature represent natural intelligence in its most primary expression, since the formation of stable stars, structured galaxies, supernova events (in which all the elements heavier than hydrogen and helium are manufactured), planetary formation, and biological evolution are all due to the consequences of natural laws, they can all be considered not only as reflections or translations of natural intelligence as I have previously called them but as *secondary expressions*. Which is to say that the physical Universe and all the processes occurring within the physical Universe can be viewed as secondary manifestations, or secondary expressions, of natural intelligence. Science, of course, is mainly concerned with documenting these secondary expressions (a possible candidate for a tertiary expression of natural intelligence might be consciousness—this idea will be dealt with in chapter 10).

Already we can divine a kind of hierarchy in which various aspects of natural intelligence are nested, or embedded, within one another. Natural laws represent natural intelligence in its primary expression and give rise to, or cause, or command, secondary expressions such as the emergence of stars, planets, and self-replicating DNA. DNA-based systems can then evolve according to natural selection. An ensuing tree of life begins to grow and develop and so on. Natural laws resolve themselves as the essential ingredient of the all-embracing environment we call Nature.

Alternatively, one might argue that the Universe just *happens* to be the way it is and the laws of Nature just *happen* to promote and sustain the evolution of life. So while life itself might be an expression of natural intelligence, this natural intelligence may not have existed in any shape or form before the advent of life. I don't know about you, but given that evolving life is in a real kind of *tutorial* relationship with its surround, I find this difficult to believe. It really depends upon how one judges the larger environment.

APPRECIATING THE ENVIRONMENT

It is surely no coincidence that we live in an era in which environmental crises of one sort or another are impressing themselves upon us. The root of the problem lies in a conceptual folly according to which we isolate ourselves from the vast contextual system in which we are rooted. Cities help propagate this self-enforced severance from the greater environment by acting as cocoons woven according to the dictates, desires, and values of human intelligence as opposed to natural intelligence. One has only to consider human culture's fervent exploitation of the planet and its resources over the past few hundred years and the environmental cost of such remorseless activity to realize that our species lacks insight into the sensibly interwoven nature of all life. The fragmentation of science into departmentalized divisions with little or no overlap also reflects this tendency to disconnect from the whole. But if we do acknowledge that life is connected into a sensible weblike matrix or pattern, then it becomes clear that this pattern extends outward and connects everything, that the entire Universe with its laws and its constants is the ultimate systemized environmental context that drives all evolutionary processes. It is surely in harmonious accordance to this larger context that we must strive to live.

It seems, then, that the environment, despite it still being a fuzzy concept (the notion of boundaries remains unclear), is nonetheless rising before us as an issue of urgent importance in more ways than one. Given man-made contributions to global warming, pollution, the destruction of species and habitats, and all manner of other environmental degradation, it is slowly dawning on our scientists and our politicians (and even a few religious leaders) that the entire biosphere really is an interconnected system within which all species and, indeed, all genes affect one another. The environment is assuredly important, and as a perceivable and conceivable system it cannot be ignored forever. That, at least, is nigh on impossible to deny these days, especially with the severe disruption of weather patterns that, perforce, everyone notices regardless of which city or which continent they live in/on.

But to really grasp natural intelligence and let it start reshaping contemporary values and behavior, it is necessary to look further afield with the mind's eye, to realize that the biosphere so rich with evolutionary activity sits within an even greater environment, that environment that we call the solar system. And that this vast contextual web is likewise a part of the entire fabric of Nature with its various laws and constants that govern the physical interactions and physical properties of everything existing. This is the ultimate way in which to conceive of the environment: as a single coherent contextual web in which all parts—including laws—are interconnected and continually influence one another. Nature is this system, and intelligence its most stunning attribute. If one conceives at this magnitude, one can see that

Nature, as a self-organizing system, is graced with intelligent structure and intelligent behavior throughout, and that the Universe might well reflect the means by which natural intelligence fulfills some kind of specific and necessary function.

That this rather remarkable assertion is true is perhaps most evident in the evolutionary process that has taken control of our planet and that has subsequently led to conscious human intelligence. Something important is surely afoot. But before we go on to speculate just what it is that natural intelligence might have in store for the future—its ultimate sense-making agenda, as it were—I should like next to focus upon some more of the reasons (apart from the backlash against religious beliefs) why natural intelligence has thus far not been acknowledged by science, and also to show how intelligence can be defined as a process. In particular, once it is made clear that intelligence, in any form, can be understood as an information-gaining process, then it can be more properly tied together with evolution, since evolution is also an information-gaining process. We look at some of these issues in the following chapter. After this follow chapters describing some of the most astonishing accomplishments of natural intelligence. Beyond that, we shall delve into the more profound implications of this new paradigm.

CHAPTER 6

ے

AH, BUT CAN NATURE PASS AN IQ TEST?

Threats to the human ego aside, the relatively slow pace of biological evolution probably represents a further reason why natural intelligence has thus far been unsung. Since we cannot generally perceive biological evolution in action—as it actually happens—evolution seems to us painstakingly sluggish and therefore, by implication, dumb. So even though we can oppose Monod's assertion that chance lies at the heart of evolution—by pointing out that ordered non-random environments provide the sense that "favored mutants" feed upon and reflect—it is still the case that evolution, subjectively speaking, takes an incredibly long time to manifest. However, this perception is indeed subjective and therefore relative. For if we took an advanced computer program—which, when running, is a process just as evolution is a process—and slowed the program right down, then such a process would also appear really dull and non-intelligent to our perception.

As an example, imagine a computer program that can recognize the human voice and converse. The kind of computer program portrayed, for instance, in the classic movie *2001: A Space Odyssey* or in the classic series *Star Trek: The Next Generation*. The program can register a series of sounds and then parse the sounds into individual words, which it can "understand" and respond to. Clearly this would be a very neat program, maybe the result of decades of research by software designers (it is no small task to analyze a *constant* audio vocal wave and tease it apart into *discrete* words). One could turn the program on, ask it some question, and find that it responds in an intelligent way. Doubtless they'll soon have such programs as part of the standard operating software for all household PCs (current software programs that recognize speech are becoming popular, although these are, as yet, not for complex conversation). In any case, if, after speaking something, we suddenly slowed down such a voice-recognition program, or algorithmic process, and then peered inside the circuitry of the computer, we might actually be able to detect individual bits changing. Bit change by discrete bit change would transpire as lengthy strings of binary information (ones and zeros) passed sluggishly through the computer's CPU. How non-intelligent can you get? If we slowed things down to a bitty crawl, it might take fully one year for the program to respond to a single uttered sentence, each hour in that year representing the yawningly slow bit-by-bit manipulation of a seemingly endless string of binary computer data. In this respect, the smartness inherent in the computer algorithm/program has all but disappeared. But if you speed the program up so that millions of bits of data are crunched in seconds flat, then we more easily note the smartness and finesse of the algorithm. We speak a sentence and, lo and behold, the program responds instantaneously. The intelligence embodied in the program now becomes very much evident.

The same is true of *all* computer programs. When running, each program is executed bit by bit, whether the program in question is some entertaining video game, an operating system like Microsoft Windows, or the software governing the robots used to put cars together. Yet these bit-by-"dumb"-bit processes, when running at their intended speed, clearly display nifty intentionality and purpose—as evinced by the specific bit changes being effected by the computer involved. When processed at their intended speed with millions of digital calculations being made each second, we realize that these discrete bit changes are not dumb and mindless at all. The same principle is also apparent in evolution. Regardless of its pace, considered in terms of what it creatively achieves and how it does so, the evolving tree of life is in no way devoid of intelligence. What really matters here is the relative way in which you observe the tree. Some views are more illuminating than others.

It is informative to see how other people deal with the slow pace of natural intelligence as it manifests within the evolution of organisms. Take the following stance promoted by popular futurist Ray Kurzweil in his now classic award-winning book *The Age of Intelligent Machines,* as well as its sequel, *The Age of Spiritual Machines.* While these books are chiefly concerned with computer technology and the future of smart machines, Kurzweil initially discusses the intelligence evinced by biological evolution. However, he quickly proceeds to eliminate natural intelligence, thereby allowing him to concentrate his attentions upon human intelligence:

> Whilst it is true that evolution has created some extraordinary designs, it is also true that it took an extremely long period of time to do so. Is the length of time required to solve a problem or create a design relevant to an evaluation of the level of an intelligence? Clearly it is. We recognize this by timing our intelligence tests. If someone can solve a problem in a few minutes, we consider that better than solving the same problem in a few hours or a few years . . . Evolution has achieved intelligent work on an extraordinary high level yet has taken an extraordinarily long period of time to do so. It is very slow. If we factor its achievements by its ponderous pace, I believe we shall find that its intelligence quotient [IQ] is only infinitesimally greater than zero.[1]

Kurzweil's reasoning can be lampooned in any number of ways. To attempt to measure the intelligence of Nature by means of an IQ test is a ridiculously anthropocentric idea, as if IQ tests were somehow the most efficient way to quantify and qualify intelligence. While an IQ test might indeed tell us something about some aspect of human intelligence (primarily pattern recognition), the domain in which such tests can be applied is strictly limited to humans. The intelligence operating within each of the billions of metabolizing cells composing Kurzweil and the long chain of evolutionary history that led to such biological wizardry can in no way be appreciated by a pen-and-paper IQ test. Such an approach is like trying to measure the internal temperature of the sun with a broken toy thermometer.

Notwithstanding these misguided attempts to gauge natural intelligence, Kurzweil's conclusion that the "ponderous pace" of biological evolution reveals a minimal IQ can be dismissed with the following example. If you gave a six-month-old child an IQ test you would not expect him or her to be able to mark answers to the questions, let alone get a single question correct. But what if, after three months of slow work, the infant, by itself and by some honest means, got every single answer correct? According to Kurzweil, we would have to curtly dismiss the results on the grounds that the child's efforts took too long to warrant the inference that intelligence was involved. Whereas most of us would conclude that the child was a prodigy of some kind and worthy of special attention, Kurzweil, according to his reasoning, would not bat an eyelid.

Clearly, then, the principle of intelligence is not to be understood solely in terms of IQ tests and other limited approaches, nor should tempo be an objective measure of intelligence. It is rather the case that intelligence, as we shall see later in this chapter, is better understood as a particular type of information-gaining process that, regardless of the time involved, is evident not just in human behavior but also in the greater processes of Nature. (Incidentally, like so many others, Kurzweil fails to connect evolution to the *entire* system of Nature. Isolating evolution from contextual considerations makes it easier to dumb down evolution.)

THE SLOW BUT SURE MARCH OF NATURAL INTELLIGENCE

As the speed of any given process is wholly relative, what really matters at the end of the day is the *specific working results,* because results will invariably betray the operation of intelligence and intent. With evolution, working results are what we should focus on, not particular fragmentary details along the way. This also applies to the kind of creative genetic algorithms we met in previous chapters. To take but one slice of a genetic algorithm's operation (a single instance of RAM state within a computer system, for example) would not highlight any intelligence or purpose in the program. Even a look at a few successive RAM states would only lead us to conclude that mechanical and entirely stupid processes (exhaustively so) were in operation. But view the program over large amounts of time (i.e., over millions of discrete RAM states) and the intention embodied in the algorithm, its specific function as deliberately substantiated by human intelligence, becomes overtly apparent, all fragmentary details blending with one another to create a *directed* result that is more than the sum of the program's otherwise discretely isolated operations. Recalling the evolution of the virtual fish eye, the algorithm used was designed specifically to do this sort of thing. The programmers did not write the software using dice and coin tosses. They designed it with the view of enabling some result, namely that the program would be able to foster the evolutionary construction of a refined virtual eye of some kind. Yet this design element and target-driven behavior all but disappear if we choose to focus solely upon fragmentary operations of the algorithm.

Having said as much, fragmentary details are, unfortunately, precisely what most of science seems obsessed with, especially when science studies evolution and biological processes. By diligently concentrating upon discrete details—like individual genes or individual mutations within DNA—science ignores the larger aspects of evolution in terms of both scale and duration (as well as context). To assert, say, that the replication of genes is the be-all and end-all of the tree of life is to ignore the greater whole, a whole that is revealed when taking into account billions upon billions of changes in DNA and the contextual environment in which those changes take place. Only when we turn our focus of attention to larger lengths of time and bigger frames of reference will we see the full raison d'être of evolution and what it is basically up to. This suggests that by dramatically speeding up the process of evolution we might get a better idea of Nature's agenda.

If we do speed up terrestrial time, a general trend of evolution toward creating more complexity and more accurate systems of sense-making becomes apparent. This is most easily recognized if we imagine speeding up the last three and half billion years of planetary/biospherical evolution into one incredibly intense minute of film. This would make a spectacular, albeit brief, cinematic movie. How would the entertainment begin? In the initial portion of this one-minute film we would witness the face of the Earth rapidly reorganizing itself in a frenzy of biochemical action as bacteria—the first organisms—materialized and then spread about the Earth, invading every possible niche and reshaping the landscape and the atmosphere. Thus, planetwide networks of nano-biology literally explode into being out of basic molecules, folding back on themselves in a maelstrom of frenetic interconnected coevolutionary activity. In the last ten seconds or so, all the animals and all the plants that have ever existed erupt into existence, the biosphere rapidly metamorphosing into an ultra-complex, throbbing, pulsating mosaic of intelligent structure and intelligent behavior. In the very last moment there is a conscious gasp of wonder.

MAKING SENSE
The Theme of Evolution

It is interesting to ponder the theme of evolution conveyed in the above cinematic thought experiment. According to this time-lapse film, the essential impulse underlying the reorganization of the Earth's surface is really quite profound. Recall that Dawkins speaks of the Universal Utility function as being the drive to replicate. Is this what most clearly emerges when imagining three and a half billion years of evolution speeded up into one volatile minute? Well, clearly replication is essential to evolution. But what is it that determines replication? Fitness. And how is fitness determined? By natural selection, by the ever-watchful contextual eye of Nature. And what is the actual key to being selected? The answer, as already noted, is the ability to make sense, for all organisms represent particular methods for making sense of the environment, of making sense of the already sensible context in which they find themselves. Which means that our time-lapse film in which three and a half billion years of evolution are speeded up into a single minute would reveal the art of sense-making made globally manifest, the entire biosphere representing an intelligent response of Nature to its own inherent intelligibility.

In a boiling, seething, frantic fervor of self-organization, elements and molecules upon the Earth's surface cohere, dance, entwine, and, ultimately, become orchestrated into enduring sensible structures and patterns that are attuned to the sensibleness prevailing all around them. From humble beginnings of basic replicating molecules, more and more complexity is realized as more and more of Nature's inherent sensibleness becomes reflected and bound up in smartly organized autopoietic architectures. The key factors here are organization, pattern, and sensible structure. They do not arise simply through chance or by brute factual luck, but because the contextual fabric of Nature—so rich to the core with intelligibility—commands such eventualities. The arising and honing of huge conglomerations of self-organizing sense-making molecules—digital DNA-writ life-forms and their constant evolution—is no less than a translation of one kind of intelligence (the logic of Nature at large) into another (localized bio-logic). The entire process might be slow—for us—but it works. And judging the process from its explicit results, evolution is clearly a clever process. Indeed, evolution does not simply work; it works actual wonders, the very real wonders of the living world. In this respect, the Earth (and any other life-bearing planets, for that matter) represents a high-resolution canvas upon which the universal presence of intelligence is reflected, recorded, and described. Slow or not, natural intelligence sets its own pace.

SEEING THE INVISIBLE

Actually, modern technology is now able to make certain aspects of otherwise invisible natural intelligence become glaringly apparent. The intelligence of our species can create technology that, in the right hands, can be employed to reveal the natural intelligence all around us. This is particularly true in the case of plant behavior. Back in the 1990s, the BBC screened a memorable six-part series called *The Private Life of Plants* hosted by revered naturalist Sir David Attenborough. The series cost millions to produce, but this was money well spent. The chief revelation, the almost psychedelic education thrust upon the viewer, was in divining the way in which plants "secretly" behave. Using the technique of time-lapse photography, certain plant behavior was made discernible, even though such behavior is normally occluded to us due, once again, to the relative speed of human perception. Plants were seen whose tendrils groped around looking for light and surfaces upon which to climb. A carnivorous pitcher plant was seen to *exude* perfectly designed pitchers (impressive jug-shaped organs for trapping insects) from the ends of its leaves. Stems and shoots were seen to purposefully waver about, tentatively looking for the right place in which to grow. Flowers were seen to follow the sun across the sky. Needless to say, some of this footage was absolutely extraordinary, revealing aspects of botanical intelligence otherwise hidden to human perception.

Consider Sir David's time-lapse-based description of the early growth of the South American rainforest cheese plant (which grows on other trees). With your mind's eye you can appreciate the artful intelligence it employs but of which we are usually unaware due to its relatively slow rate of execution:

> The seeds of a cheese plant are no bigger than orange pips . . . When they ripen and fall, they scatter widely over the forest floor and almost immediately germinate. Green worm-like shoots slowly writhe out of them and begin to extend across the ground, heading for the base of the tree from which they have tumbled. If a great number of them have fallen and they are evenly distributed over the ground, then as the shoots grow they begin to look like the spokes of a huge wheel, the hub of which is the bole of the tree that still supports their parent. It seems almost miraculous that they should all, in some way, know where to go. They do because, like all plants, they can sense the light. But they, unlike most shoots, do not seek it. They are programmed to avoid it and they head for the nearest deep shade. If their parent was sitting on the far end of a branch, then they may move toward the base of a neighboring tree. Failing that, they will creep toward the one from which they have fallen . . .
>
> If they fail to find a tree trunk within six feet or so, then they run out of fuel and die, exhausted. But if one encounters a vertical surface within that distance, it suddenly changes. Instead of seeking shade, it seeks the light. It begins to climb upward. Small round leaves spring out from either side of it and they . . . produce food. With this new fuel supply, it climbs more strongly. As it ascends, its leaves get bigger. By the time it is approaching the canopy, where the light is brighter, its leaves, which are now a foot or so across, begin to divide into segments . . . A single cheese plant produces . . . three kinds of very differently shaped leaves, each suited to a different circumstance and phase in its life.[2]

All in all, after watching the extensive time-lapse footage shown in this amazing series, one was left with the impression that one had witnessed some alien species of life, that plants were a bit like "triffids," highly intelligent creatures with a repertoire of uncannily purposeful behavior that is normally hidden to us because of its relatively slow speed. But with the advent of special filming techniques, this slowly manifesting vegetal brand of intelligence becomes very much apparent. Plant organisms are quite literally morphing in clever ways all the time.

Perhaps some bright spark will eventually attempt to make a computer animation called *The Private Life of the Earth,* thereby creating a version of the film I asked you to imagine earlier. This would surely convey the notion of natural intelligence even more dramatically than Sir David's film or our purely imagined film. It would be nigh on impossible not to divine the operation of an organic intelligence upon viewing the instant bursting out of life across the face of the globe. And it is not that time-lapse techniques are misleading in any way. Quite the contrary. What time-lapse footage can do is fade out the lesser "melodies" and "undertones" of a process and bring to the fore the most prominent and essential themes. To view *The Private Life of the Earth* would be to behold the birth and elaboration of a planetary membrane, a diverse yet singularly coherent biotechnological skin evincing more engineering prowess than anything we humans have ever crafted.

EVOLUTION VIEWED AS AN INTELLIGENT PROCESS

Thus far, the assumption has been that intelligence can be understood as a particular kind of process that is not necessarily tied to human minds alone (and that it need not be conscious). Hence my contention that intelligence is a property of Nature, being most clearly manifest through biological evolution along with the laws that foster and nourish biological evolution. Regarding the intelligence inherent in both the laws of Nature and the software laws governing genetic algorithms, while such laws are not themselves processes, they give rise to intelligent processes, and in that sense laws too can be considered as embodying intelligence. However, for the purposes of what follows, we shall concentrate on intelligent processes alone and not the more abstract and seemingly static laws that foster them.

While the assertion that Nature is inherently intelligent may seem sound to some readers, and while it may be a compelling way to account for the theme of evolution, other readers will doubtless still feel uncomfortable with stretching the definition of intelligence. After all, we generally associate intelligence—or at least advanced intelligence—solely with humans. Like Kurzweil, we think of it as a thing, a kind of object, or noun, that can be readily quantified, for instance, by intelligence tests. We speak of humans possessing more or less intelligence and so on. It is less often that we talk of biology or organisms as being intelligent, let alone the whole of Nature. So, when you get down to it, what exactly is intelligence? And can intelligence really be located in systems different from the brain? Here too may lie another reason for the negation of natural intelligence. Unlike the last reason—the relatively slow rate of evolution—this reason has more to do with our concepts about what constitutes intelligence.

The suggestion presented here is that intelligence is a sense-making process, or sense-making capacity, that can manifest in different systems apart from the human brain (or chimpanzee brains or dolphin brains). After all, the sciences of artificial life and artificial intelligence already mentioned in these pages pay homage to the fact that intelligence can indeed be embodied within systems like computerized robots and digitally evolved computer programs. Moreover, artificial intelligence is obviously a well-established science, since we are all familiar with it to some extent, unlike the complementary notion of natural intelligence. Here we have a disparity of great importance. If we can get to grips with what artificial intelligence represents—why it is so-called—then we ought to be able to see more clearly what natural intelligence represents. This is really no different from the manner in which Darwin compared and contrasted artificial selection with natural selection. The only real difference is that natural selection has been accepted, whereas natural intelligence has been denied outright.

ARTIFICIAL INTELLIGENCE (AI)

The intelligence being created by AI scientists is bound up with the behavior of the machines they build. John McCarthy, the founding father of AI, originally defined the new discipline as "the science and engineering of making intelligent machines." Thus, AI scientists design and build machines that can run on their own and behave in a sensible and intelligent way. Take, for example, NASA's illustrious 1997 Viking Pathfinder mission to Mars in which, to the accompaniment of howls of joy from NASA staff, the small AI rover vehicle (called *Sojourner*) rolled smoothly out of the landing ship and onto the red Martian surface in order to analyze rock samples and take pictures. It was one small drive for the rover, but one giant leap for AI robotics. The rover was truly a triumph of the AI community since its onboard computer was able to navigate safely across completely unfamiliar terrain. In this sense the rover wielded a certain degree of intelligence that was manifest in its non-random structure and in the non-random behavior it executed according to its onboard computations (since 2003, two more rovers have been sent to explore the surface of Mars).

To give the reader some idea of the sophistication of the pioneering Mars rover vehicle, note that it took three and a half years of concerted effort to design its computer hardware and software. The purpose of this computer system was to give the rover a fair degree of *autonomy*. Indeed, autonomy—the ability to survive and function without direct intervention—can be considered one of the most marked attributes of AI robots, a clear sign of their inherent intelligence. Given that radio signals from Mars to Earth can be delayed by ten minutes or more, it obviously made more sense to design the rover with some degree of autonomy than to remote-control it from Earth all the time; hence the partial autonomy, or artificial intelligence, embodied within the rover. Equipped in this fashion, the rover was able to carry out many operations on its own. More than a hundred million miles from home, in a strange new world where no such robot had been before, the diminutive rover was nonetheless able to explore and record data about the red planet. Fifteen years on, it's still up there, like an eager and faithful hound, waiting stock-still for new instructions, which, sadly, will never materialize.

As an example of the rover's hardwired intelligence, its onboard computer would run a hazard detection program. This allowed the vehicle to identify any obstacles en route to whatever destination the ground control team directed it. Using two cameras and five lasers (neat body parts), the rover was able to analyze its surroundings and then calculate whether any obstacle in its way could be traversed or not. If such an obstacle were computed to be too big to traverse, the rover would promptly calculate a route around it. Similarly, if the ground team, in some drunken fit, were to command the rover to drive off a cliff, it would disobey, according to the overriding dictates of its autonomous hazard-protection system. In layman's terms, you could not mess with the rover and it did not take any nonsense.

The rover also issued what was known as a "heartbeat." It would frequently stop and communicate with the lander vehicle, transmitting its current location. If it veered too far away from the lander for its radio signals to work, then it would actually retrace its steps to a point where communications could be reestablished. This is reminiscent of the way an infant animal will stay within earshot of its parental keepers. And the ability of the rover to retrace its steps—by recording and reversing wheel revolutions and such—recalls the ability of social insects to return to their nest.

Another aspect of the rover indicative of its artificially embodied intelligence was a certain thermally insulated box device inside it. Held within this box were electrical components unable to function at the bitterly freezing temperatures that prevail during Martian nights. The box effectively maintained a stable working temperature in which the electrical components could be preserved and function adequately. This brings to mind the homeostatic heat regulation properties of living organisms that likewise serve to protect organisms from severe changes in temperature.

What AI engineers do, then, is design robots whose physical structure *makes sense* in the contextual environment in which they are to be employed and whose onboard computer systems can make *moment-by-moment sense* of their environment (somewhat analogous to thinking). Once they can make sense, they can navigate and do sensible things in that environment (as well as survive, of course).

In order to make moment-by-moment sense, the computing systems housed by these robots need to absorb information, store it, and then, where possible, use that information to adapt and learn. The Mars rover vehicle was able to learn to a certain degree, since, if it confronted some novel large object, it could compute, or learn, how to circumnavigate the obstruction. The potential for learning and making sense obviously depends upon the complexity of the programs employed by a robot's computer system. If the programs are complex enough, or utilize genetic algorithms, then their capacity to learn and make sense increases.

THE ESSENTIAL CHARACTERISTICS OF INTELLIGENCE

These four items—the absorption of information, the storing of information, the ability to learn from that information, and the ability to make sense—are, I contend, the principle ingredients of intelligence. Indeed, they are precisely the sorts of things that we do so well and that the robots of AI engineers attempt to do. Our senses continually take in information about the environment, our brains store this information, and then we use the information to learn and to make ever more sense of the environment. These are all clearly behavioral processes, not static things. Bearing this in mind, it should be apparent that the end result of these four processes working in combination is intelligence.

Intelligence is all these processes combined into one, the term "intelligence" being a convenient way of explaining their combined action. So, intelligence is definitely not a thing in the way that a table is a thing. Intelligence reveals itself as a process, moreover a process concerned with the *constructive manipulation of information.* The science of AI is all about substantiating this process in robotic form, or *in silico* as some AI scientists call it.

There is nothing unique about this definition of intelligence. It is not controversial or radical. It simply sees intelligence as a process concerned primarily with the non-random manipulation of information and not as some abstract thing. Such a process can be embodied in different media. Indeed, even a thermostat—a cybernetic device dependent upon feedback—can be viewed as a very primitive example of manifest intelligence (or at least proto-intelligence), since its parts serve to monitor the environment and act accordingly by switching other machinery on or off. In a real way, a thermostat processes information and uses it to perform a useful function. What is important in a thermostat is the (cybernetic) arrangement of its information-sensitive parts and what those parts, in total, achieve.

We humans are intelligent—more so than thermostats, of course. Our five senses pick up vast arrays of information and convey this information to the brain. Somehow, the brain organizes this information and allows us to make use of such information by learning. It might even be that a process analogous to natural selection takes place in the brain through neuronal pathway pruning of some kind (so-called neural Darwinism). In the same way that natural selection prunes the growing branch tips of the tree of life, so too might some similar pruning process be constantly occurring within the neuronal pathways of the cortex (maybe even at a quantum level). Learning, indeed all forms of learning and sense-making, would then be synonymous with the non-random pruning of various "informational possibilities" according to some enveloping context. In any case, the crucial thing about learning is that it allows more information to be accessed, more certainty to be attained, and more sense to be made. And there it is again, the notion of sense-making. Already we can understand that sense-making really is bound up with intelligence. The smarter or more intelligent something is, the more sense it can make. Modern humans—*Homo sapiens*—are presumably smarter and able to make more sense of the world than our hominid ancestors millions of years ago. Likewise, NASA's latest Martian rovers are presumably smarter and more capable of sense-making than the original *Sojourner*.

THE INTELLIGENCE OF EVOLUTION

If one accepts that intelligence involves the accessing, storing, and re-organization of information such that sense can be made and learning can occur, then evolution must assuredly be considered an intelligent process, as this is *precisely what evolution is all about*. The information that is accessed through evolution is the lawful a priori sensibleness or meaningfulness of Nature itself (being extremely ordered, Nature is replete with specific patterns of information). The storage or encoding of this environmental information takes place in digital DNA. The reorganization of such information takes place through the non-random selective variation of that DNA. As Richard Dawkins himself has conceded: "Natural selection is *by definition* a process whereby information [about the environment] is fed into the gene pool of the next generation."[3]

Considered over large spans of time, and considered as a process involving the gaining of specific information, evolution is therefore (naturally) intelligent, with organisms and their repertoire of genetically based bio-logic representing learned ways of being, learned ways of making sense, learned ways of surviving.

Actually, we can go further and state that bio-logic represents a natural form of "understanding" or even "awareness" of some kind. For instance, everything that a cell does testifies to the fact that it "understands" the laws of chemistry in the same way that a clever chemist understands the laws of chemistry (although, of course, a cell's intelligence need not be conscious). Both are able to exploit the sensibleness of chemical law to their advantage. However, given that cells have perfected metabolism (an *extremely* smart process), they can be said to embody a greater understanding of, say, the micro-world of amino acids, proteins, lipids, and sugars than a human chemist of which they might be component parts. The point is that sense-making, learning, and understanding—all indubitably bound up with intelligence—are exactly what evolution implements via bio-logic. Here are Lynn Margulis and Dorion Sagan once more, this time commenting on this wholly natural learning process:

> Problem solving began 4,000 million years ago by way of mutating, recombining prokaryotic DNA chains. Natural selection, by preserving the bacteria and their descendants with the most effective responses to the environment, stored solutions to problems of overheating, drought, and ultraviolet radiation. The form of storage was as informative sequences of nucleic acids.[4]

Take a more vivid example, that of flight. It is usually thought that animals attained flight step by step from some primitive semblance of flying. Flying squirrels come to mind. Utilizing taut winglike flaps of membrane, a flying squirrel with outstretched legs can effectively glide from tree to tree. Moving our attention to the evolution of flight by birds, if we imagine that a linear evolutionary sequence of, say, a hundred ancestral bird-type species (i.e., reptiles, or even dinosaurs) were involved in the evolution of proper full-blown flight, each successive species will obviously be more physically adept at flight. Each of the one hundred successive species therefore embodies one closer step toward the realization of refined flight. The entire sequential process can be seen as a learning process. Species, or genera, quite literally learn to fly over time. Through successive instances of natural selection, bio-logic effectively learns the mathematical and engineering wisdom necessary to implement the precise muscle structure, muscle coordination, and wing design required to achieve heavier-than-air flight. In more dramatic terms, one can say that avian wing structure and avian musculature embody an acute "understanding" of aerodynamics. If, however, you focus your attention on but one species in this sequence of one hundred, you will not divine the learning. But if you focus on the entire process, the sequential progression so to speak, a naturally intelligent learning process becomes apparent (albeit unconscious). To be sure, *all* evolutionary innovations can be viewed in the same way, whether we think of locomotion, sight, metabolism, or any other biologically determined behavior. In every case, a natural process of learning is synonymous with evolution.

It should hopefully be clear by now that high-level intelligence—the competent accessing, storage, and reorganization of information—is not a phenomenon associated solely with the human brain/mind complex. Indeed, the actual human brain—its structure, organization, and capacity—is *itself* a result of an aeonic intelligent learning process undertaken by evolutionary bio-logic. The wisdom so accrued, the sensible solutions to being in the world, is found in the DNA of the cells that compose organisms, the genome of any organism representing a kind of encyclopedia of knowledge and intelligent "understanding," parts of which, in the human genome, code for the actual human brain. This is all part and parcel of natural intelligence, which dictates both that evolution happens, and also that evolution drives the formation of more and more refined forms of biological structure and biological behavior. Human intelligence can, therefore, be understood as an elaborate *by-product* of natural intelligence, the latter having constructed the brains that allow for human intelligence. Natural intelligence is thus more fundamental and vastly superior to human intelligence. An ego-shattering conclusion once again, but such is the nature of the new paradigm.

In summary, evolution can be interpreted as being a naturally intelligent process operating over large amounts of time. Evolution, via the language of bio-logic, learns the art of sense-making, of making sense of an already sensible context, this context being the lawful and ordered system of Nature. Such learned sense is written down in the form of enduring and heritable DNA. And the plasticity of this DNA, its tendency to vary and change over time, ensures that the learning, the sense-making, continues, thereby driving the evolution of smarter and smarter systems of bio-logic. Eventually, this creative state of natural affairs led to the refined human cortex able to marvel over its origins. It matters not that the constructive sense-making process of evolution manifests over billions of years. What really matters is the resultant tree of life, each speciated part of which represents a rich embodiment of intelligence. Not only has the tree of life been planted, watered, and fed by natural intelligence; it is itself a reflection of that self-same intelligence, embodying it in each of its component parts.

ARTIFICIAL INTELLIGENCE VERSUS NATURAL INTELLIGENCE

We have seen that the study of artificial intelligence testifies to the fact that intelligence, as an information-gaining process, can be embodied in robots. What has not been expressly stated is that AI scientists actually borrow from natural intelligence. Or perhaps the word "plagiarize" is more appropriate, given that natural intelligence is not explicitly acknowledged by most AI scientists. As stated, natural intelligence (or NI) is unheard of. While the AI fraternity might not admit to an accusation of plagiarism, it is undoubtedly true. For instance, AI scientists interested in building flying robots will study the flight of insects and birds in order to get inspiration. Likewise, AI scientists interested in building underwater robots analyze the movement of fish and dolphins. Those keen to develop nanotechnological (i.e., ultra-small) robots observe bacteria and cellular processes that are themselves working examples of the nanotechnological inventions of natural intelligence.

If you think about it, Nature is bound to provide the best examples of the skills we are keen to explore and re-create in artificial form. Having had hundreds of millions of years to hone themselves, species of life like fish and birds are expert in making sense of the environments of water and air, respectively. The sleek shape of a shark, for example, its graceful body-rippling method of locomotion, likely represents the most efficient and sensible way of moving through water, the mathematics for which are presumably buried somewhere in the shark's genes. The same goes for the ability of birds to fly. Not only have they perfected the art of heavier-than-air flight, it is likely that had we not witnessed the flight of birds (and insects), then we would never have dreamed of building flying machines.

This important insight—that Nature is ultra-smart and the ultimate design force—has been noted by Kevin Kelly, whom we briefly met a few chapters ago. He writes:

> Life is the ultimate technology. Machine technology is a temporary surrogate for life technology. As we improve our machines they will become more organic, more biological, more like life, because life is the best technology for living . . . What we know as life today will remain the ultimate technology because of its autonomy—it goes by itself, and more importantly, it learns by itself.[5]

The same situation holds true for the booming interest in neural networks among AI scientists. A neural network represents a smart control-system architecture digitally substantiated within a computer system, a kind of artificial thinking system. Any robot or computer able to carry out really clever tasks will be sure to be in possession of a neural network or two (that's why you hear of them in so many sci-fi films). Once again, the idea of the neural network arose from observations of natural intelligence. As artificial life expert T. S. Ray explains, with reference, I should add, to many AI phenomena we have already met:

> One of the greatest challenges in the field of computer science is to produce computer systems that are "intelligent" in some way. This might involve, for example, the creation of a system for the guidance

of a robot that is capable of moving freely in a complex environment, seeking, recognizing, and manipulating a variety of objects. It might involve the creation of a system capable of communicating with humans in natural spoken human language, or of translating between human languages.

It has been observed that natural systems with these capabilities are controlled by nervous systems consisting of large numbers of neurons interconnected by axons and dendrites. By borrowing from nature, a great deal of work has gone into setting up "neural networks" in computers. In these systems, a collection of simulated "neurons" are created and then connected so that they can pass messages. The learning that takes place is accomplished by adjusting the "weights" of the connections.[6]

More evidence, then, to support the paradigm of natural intelligence. To reiterate: intelligence of one kind begets intelligence of another kind. If we are to construct extraordinarily advanced robots and cognitive computer systems as envisaged in sci-fi films, then we have to tap in to the wiles and ways of natural intelligence, the prime mover, which has already achieved, in biological form, the sort of thing human engineers dream of achieving. Man always learns from Nature, because Nature is the ultimate source of all manifest intelligence.

Thankfully this notion of Nature's engineering acumen is nowadays becoming acknowledged and championed by the burgeoning biomimicry movement. Spearheaded by the ever enthusiastic Janine Benyus, biomimicry is often described as being "the conscious emulation of Nature's genius." In other words, the conscious emulation of natural intelligence (even though "natural intelligence" is not spoken of as such). All manner of natural designs are being investigated and copied—anything from the ventilation systems of termites to the self-cleaning capacity of lotus leaves to the underwater glues utilized by the barnacles we met in an earlier chapter. The assumption is that Nature has had millions of years of research and development, and that we would therefore do well to learn from Nature. Indeed, given that life on Earth has persisted for over three and a half billion years, we can learn from Nature not only hard-won engineering techniques but also the art of sustainable living. As far as I can see, the biomimicry movement heralds a much needed new vision of Nature, one in which human intelligence is brought down a few notches and begins to appreciate that the larger web of life is smarter than we have previously imagined.

PUTTING ARTIFICIAL INTELLIGENCE IN ITS RIGHTFUL PLACE

Like biotechnology, AI is a prestigious industry, considered the cutting edge of human innovation and human invention. In various labs throughout the world—especially in the productive and heavily funded labs of the Massachusetts Institute of Technology (MIT)—engineers and computer scientists are busy building more and more complex robots that are more and more efficient at executing intelligent behavior. Many of these robots resemble oversized insects. They often possess six legs in the manner of insects, and this allows them to maneuver steadfastly across bumpy terrain. Indeed, AI engineers have realized that the characteristic "tripod gait" of insects, in which three of their six legs are on the ground at any one time, represents a maximally stable approach to terrestrial walking, and this is why robots are built to mimic this neat technique. The software (including neural networks) running on these insectile robots is generally able to facilitate the robots' movements and goal-orientated behavior. Certain objects can be steered toward, other objects avoided, and so on.

These modern insectile robots echo the earliest days of AI. A pioneering scientist once designed a light-seeking robot on wheels. If its onboard sensors registered more light coming from one direction, a signal would be sent to its wheels such that some of them would stop rotating. This meant that the robot would swivel around. It could then proceed more directly toward the light, stop, alter its wheel spins again, and then continue. In this fashion it would eventually arrive at the source of the light, perhaps having its solar batteries recharged. This early AI experiment shows how a small slice of intelligent and purposeful behavior can be molded into a wheeled robotic machine.

The robots they build today, such as Mars rovers, are of the same kind, although more refined, more versatile, more capable of processing information and learning. But they are never as smart as living organisms. Far from it. Indeed, the robots produced by the AI community unwittingly highlight the design superiority of living creatures that robots seem wont to mimic. In other words, natural intelligence excels, whereas artificial intelligence gropes and mimics. As M. G. Dyer, an artificial life scientist, attests:

> While the nature or even existence of conscious thought is highly controversial in animals, it is clear that the tasks performed by animals require enormous amounts of computation. For example, consider cooperative hunting in carnivores or nesting site selection and group nest construction by bees, weaverbirds, or, among mammals, the dams built by beavers. The sensing, locomotive, manipulative, and social skills exhibited by such animals in performing these tasks completely eclipse any kind of robotic behavior produced so far within the field of AI.[7]

SIZING UP THE COMPETITION

Let's take this comparison between artificial intelligence and natural intelligence a little further. Take the labs at MIT where they design the latest AI machines. Imagine that there is some competition being run, a large prize to be awarded to the MIT team that can produce the most sophisticated all-purpose robot. All sorts of prospective plans would be drawn up. Various engineers and designers would get together and start coming up with ideas. Technical drawings would be passed around and mulled over. A stage would come when early prototypes were made and evaluated and so on.

Now, imagine that we burst into one of the labs planning to win the competition. We proudly place on the lab table a small dark and typically insectile device, a kind of miniature version of Genghis, a very famous foot-long cockroach-shaped robot developed by MIT in the early 1990s. We say that we have called the device Bug. We casually explain that our bantam version of Genghis, Bug, can walk over any terrain. Up walls if necessary. Even upside down. We explain that Bug can also right itself if it finds itself toppled over. We then declare that it can *fly* too. This would guarantee gasps of amazement from the AI scientists. They would peer more closely as we indicated the lightweight wing devices connected to Bug's insectile outer casing. To further murmurs of amazement, we point out that Bug does not use lithium batteries or solar panels. In fact, we proudly announce that Bug's energy source is derived from cheap and readily available organic plant detritus; by an ingenious hardwired behavioral routine, it can collect and convey plant matter into a special internal compartment where the energy in the matter can be turned into locomotive and computational power.

Someone asks us the life span of Bug, hoping that at least in that department our device is limited. We explain, however, that Bug was actually built *one year ago* and is still going strong due to *self-repair mechanisms*. By now, heads would be shaking with disbelief. It would seem like too much, that Bug was simply far too amazing to be true. But we have more features to extol. We explain that Bug, mounted with a microscopic camera and microphone, can be sent on reconnaissance missions into crevices or other places where humans find it difficult to penetrate—during, say, the aftermath of an earthquake when buried victims need to be located and rescued.

No doubt at this point some of these AI scientists will be baying for information on exactly how Bug was designed. Did we collaborate with Japanese AI scientists? Were we sponsored by some wealthy organization like Apple, IBM, or Sony? Did we invent some new kind of neural net? Did we raid and steal plans from futurist Ray Kurzweil's secret underground laboratory? Did we utilize a multitude of complex nanotechnological components? Did we base the design of Bug on a secret new-generation Mars rover vehicle perhaps? And so on. But before we begin to address these questions, we have yet more astonishing features to describe. We now boldly inform these AI scientists that a good number of Bugs exist and that in certain situations they can meet and exchange complete software programs pertaining to how they were built. In the right conditions, these lengthy codes can combine and actually cause the formation of more Bugs, built from basic environmental molecules and not from synthesized silicon and aluminum. In other words, Bugs can replicate on their own without human intervention!

Well, our AI scientists will surely have started to swoon and faint by now. To cap all these features of Bug with the assertion that the thing can verily replicate on its own would seem outrageous in the extreme. Certainly Bug would be guaranteed to win the competition legs down. Indeed, knowing they have not a hope in hell of matching the magnitude of intelligence inherent in Bug, the AI scientists at MIT would probably promptly give up on their own inventions and prototypes and beg us to share in the obviously advanced technology embodied in our device. Bug would be the envy of every AI scientist in town, its autonomy and intelligent repertoire of behavior being leaps and bounds ahead of anything previously designed. And to think, we told these AI scientists about only a *fraction* of Bug's innate intelligence . . .

REAL BUGS

Of course, the above scenario is a parody designed to highlight natural intelligence. Bug really *was* a bug, a real living black beetle to be precise, sequestered, let's say, just outside the labs at MIT, where it was stomping through the undergrowth busy doing whatever it is that beetles do. All the key features of intelligence so compelling to AI researchers are embodied in insects like beetles. Like any organism, beetles exemplify the art of natural intelligence and reveal that the science of artificial intelligence is not capable of matching the exquisitely perfected systems of Nature. All we humans can do is strive to learn from natural intelligence, learn its tricks and such, and then seek to embody such artistry in a severely limited fashion in some robot or computer. The lesson is a humble one. Natural intelligence is mighty, and we mill and tinker about in its extensive wake.

Yet consider the following statement from Marvin Minsky, a researcher at MIT who has been at the forefront of AI for more than thirty years. Like most other scientists, Minsky candidly asserts that "Darwinian evolution is dumb learning."[8]

Given the field of AI's reliance on mimicking the capacities of organisms, of learning from bio-logic in order to further the technology of AI, it is bewildering that scientists like Minsky steadfastly believe that evolution, and by association, Nature, is lacking in intelligence. Minsky would have been better to replace the word "dumb" with "natural"—at least then he would have been closer to the truth. After all, Minsky's cortex arose through evolution (recall that the human cortex is said to represent the most complex device in the known Universe). Yet, just as in the case for Dawkins and Dennett, Minsky's cortex employs itself so as to deny that it has been intelligently crafted: an intelligent system decreeing that it has dumb roots, somehow ignoring the intelligently configured context of Nature that ensured its evolution. If I had my way I would strap anyone of Minsky, Dawkins, or Dennett's persuasion into a chair just like protagonist Alex in *A Clockwork Orange* and force them to watch, say, the latest computer animations of genetic processes that show DNA being furiously transcribed so as to build mind-bogglingly exquisite protein machinery (the Walter and Eliza Hall Institute of Medical Research has produced some extremely impressive computer animations along these lines).* Such films are *awesome,* as if one were witnessing some uncanny alien technology at work. Advanced nanotechnology is an understatement. Indeed, even natural intelligence is an understatement! And yet how would Minsky, Dawkins, or Dennett describe the life-affirming genetic wizardry occurring in each of their cells? In what terms would they best appraise the highly specific behavior undertaken by bio-logic and so clearly evident in the latest animations of cellular processes? A triumph of dumb designoid tinkering? A masterful spectacle of feeble mindlessness? A spectacular molecular show scripted by an idiot? *Are we really to infer that more intelligence is involved in the making of such animations than in the actual processes being modeled?*

Ironically enough, it would seem that natural intelligence is so extensive, so immense, so *slick,* that it remains all but invisible even to those considered to be highly intelligent people. I leave it to the reader to decide whether this results from ego once again or from some other strange peculiarity of the human psyche (maybe because we tend to readily see parts but not whole systems). Whatever the case, it seems high time that natural intelligence be discussed and delineated with the same sort of fervor as artificial intelligence. To apprehend natural intelligence, to grasp it and ponder it, is to completely alter one's relationship to life and the Universe at large. Everything begins to change.

---

*See www.wehi.edu.au/education/wehitv.

CHAPTER 7

CLOSE ENCOUNTERS OF THE NI KIND

We have established that biological evolution is a process whereby information is gained, learning is achieved, and sense is made—exquisitely so. Evolution is therefore, by any standard, a naturally intelligent process. The source of all the learning achieved by evolving life and the sole reason that sense can be made is the larger system of Nature. It is the larger system, or context, of Nature that evokes, provokes, and drives biological evolution— by providing learning material, just like a book provides learning material. The overall result of this ongoing learning process is the spectacular tree of life whose myriad parts—organisms—constitute the established reflections, or reverberations, of the original "background" intelligence that we call Nature. "Reflection," used on a few occasions now, is certainly an appropriate term to pinpoint the ways in which Nature's inherent intelligence flows and makes itself known. If something is a reflection (or reverberation, or translation), it preserves and reproduces a salient quality of something else. Reflections and reverberations imply a significant correspondence and/or equivalence. In a novel, for instance, an author's personality may be reflected in various characters. The behavior of the chief protagonist might be a reflection of the author's innermost desires and so on. Similarly, music can be a reflection of powerful emotions felt by the composer.

The reason the term "reflection" is particularly apt when talking of organisms is that the sense that organisms make and put into practice derives from the greater system of which they are component parts. The intelligent configuration of bio-logic making up any organism can be seen as reflecting the equally intelligent configuration of the natural world. The same principle of reflection holds true with the human intelligence involved with genetic algorithms. Human intelligence is evident in both the software laws governing the algorithm and the products the algorithm generates, especially if these are products that the programmer had in mind. In both cases, the real world and the artificial world, intelligence flows and reflects itself in different ways and by different means. But even though intelligence is apparent in the evolutionary process—its essential learning capacity, as it were—as well as in the actual laws that serve to promote evolution, it is far more discernible in the actualized living, breathing products of evolution. After all, organisms and their constituent legacies of bio-logic are easier to perceive and contemplate than are the forces and laws that nourished their arising. Given that this is the case, we will now take a closer look at the impressive natural design of various manifestations of life in order to help establish the natural intelligence paradigm.

While there are endless dazzling examples of natural intelligence in organismic action—namely any of the millions of species of organisms, that share the biosphere with us—I have endeavored to find some of the more memorable ones. Let us start with the good old flu virus. Am I serious? As a prime example of an embodiment of natural intelligence? Is this bathos? No, actually, for as we shall see, although a flu virus is not a very enjoyable aspect of life, as a very basic entity it embodies sensible strategies for "living" and surviving (i.e., making sense of the environmental context in which it exists). To be sure, it is a moot point as to whether a virus is definable as an actual living organism, which explains my use of quotation marks. But this makes the following observations more marked since even such a relatively elementary quasi-living thing as a virus still embodies a fair whack of hardwired intelligence.

A flu virus is a cruel but devilishly smart piece of natural handiwork. We are surrounded by such viruses, since they fill the air in the same way as pollen grains and fungal spores. There will probably be a good few hovering above this page. Small and invisible to the naked eye (they were not physically described until the advent of electron microscopy), viruses are quite simply "naked" portions of digital DNA packed into a protective shell, rather like free-floating bits of natural computer code (they are thought to have originated as "loose bits" of DNA that went AWOL). The thing to bear in mind here is that different viruses are coded to animate themselves upon reaching different targets. Which is to say that viruses don't really do anything until they fall into the right context.

If certain flu viruses are breathed in and happen to find themselves at the back of a human nasal passage—their prime target—then they go into action, their DNA code suddenly turning on. The flu virus, having sensed the target to which it is programmed to respond, stealthily infiltrates the membrane of the nasal cavity at the back of throat. The virus then diligently heads for the insides of the cells, a mean act of breaking and entering if ever there was one. Once installed in human host cells, the virus then slavishly usurps them into manufacturing more flu viruses. This is a bit like a heartless gun manufacturer taking over an automated toy factory and making the factory churn out machine guns instead of fluffy bunnies.

But not for nothing is the human organism regarded as a robust self-sustaining piece of natural craftsmanship. In response to the vicious viral attack, the human immune system goes into overdrive in an effort to stave off these marauding microscopic parasites. Blood vessels in our nasal cavities and our throat dilate as more healing blood (full of antibodies and other defense measures) rushes to the crime scene. An unfortunate side effect of this is that our noses run and our throats swell. Copious amounts of mucus are released, which, along with the throat swelling, causes us to sneeze and cough—all incidental features of our immune system's response to the presence of the viral miscreants.

Now, when we start to sneeze and cough, we might inadvertently pass on the virus to someone else in our vicinity. That's why it is so very rude to sneeze and cough without covering your mouth. For, attached to minuscule airborne globules of mucus and saliva, newly hatched flu viruses can find themselves being transported in a rather splendid manner. What could be better for the virus? We socializing hosts actually give the viruses a free lift, sending showers of them out in all directions. Short of placing the virus directly inside another person using a long thin spatula, by coughing and sneezing all over the place we are giving the virus a free ticket to ride straight into the noses and throats of other potential victims. For the virus, of course, it is a highly efficient way of replicating itself, of finding itself in exactly the right place where it can once again wreak more copies of itself. Viruses are the professional terrorists of the organic world, causing misery and mayhem whenever they happen to waft to their preferred locale.

This side effect of our immune systems—coughing and sneezing—which ensures that we spread the virus, is, for the virus, an eminently sensible consequence to be subjected to. *And this is precisely why these flu viruses flourish.* Flu viruses contain DNA that instructs them to go into action when they are in the nasal passage/upper throat of a human—that's the exact context to which they are sensitive, their preferred target. Flu viruses are programmed to infect *only* in this precise way and *only* at this precise location. And the reason these viruses do better than a hypothetical virus that is primed to go into action when it enters, say, a human belly button or a human ear is that the side effect of our immune system dealing with a virus that targets the back of the throat is to unwittingly facilitate the replication of the virus. So although flu viruses do not *directly* make us cough and sneeze, those viruses, which happen to explicitly target nasal passages, still serve, willy-nilly, to cause coughing and sneezing, and this is why they have been selected by Nature. Flu viruses make simple but effective sense when they happen to land in the locale to which they are attuned. That's what they do. They don't know pity, they don't feel remorse, and *they will not stop* until they have done their dastardly "influenzial" deeds.

In short, then, the particular target-orientated DNA instructions within the flu virus and other prevalent viruses embody strikingly effective portions of natural intelligence, learned via trial and error over millions of years. Although considered the most elementary type of organic entity carrying DNA, a virus still manages to make sense of the world, its quota of DNA representing a *learned technique* for surviving and replicating. Simpler even than bacteria, viruses doubtless represent the smallest architectures of organically embodied intelligence with which we are familiar. Worth thinking about the next time you get a cold. It might be a pain to do battle with a flu virus, but in the sneezes and coughs you are forced to emit, you can salvage some measure of insight, namely that the virus you are harboring is testimony to the cunning intelligence encoded in its DNA (in the long run, of course, a virus could potentially *make much more sense* if it were to confer some survival benefit to the host while it was replicating itself—so maybe the "good luck virus" featured in the comedy sci-fi TV series *Red Dwarf* is a possibility, given enough time).

THE DREADED CASH MACHINE VIRUS

To drive home this point that viruses are primitive insular chunks of natural intelligence, let's imagine some new virus, say a virus that, through an unusual mutation, became attracted to the buttons of cash-dispensing machines. It *could* happen given enough time, especially when you consider the consistent ability of viruses to mutate (this explains why we keep getting colds—the cause lies with newly evolved strains of flu virus to which we have not yet developed antibodies). If such a strange and novel virus were to alight upon a button on a cash machine, its DNA could switch into action and ensure that the virus hung on at such a geometrically precise location. Which is to say that a button on a cash machine will represent the virus's initial target. Maybe the square shape of the cash button attracts it, or even the skin residue on its surface. In any case, the virus might transfer itself onto the fingertip of a human who presses the button and then invade the finger just under the nail, or something along those lines. And then of course, after having replicated like mad in the victim's finger and emerged en masse, the virus might be transferred to *another* cash machine where it could infect someone else. And so on.

Now let's extend this peculiar thought experiment. Imagine that such a cash machine virus began to flourish. Imagine that variants and mutant viruses arose with a particular *preference* for certain of the buttons on cash machines. Let's say that some variants became sensitive to the geometrically inscribed legends on the buttons, developing a preference for the green Enter button. We'll call them the "Enter Button Viruses." Since the Enter button is the button most commonly used on cash machines (whatever numbers you key in, you are compelled to hit the Enter button afterward), then those particular variant viruses would be more likely to find themselves transported onto fingers. It logically follows that these particular Enter Button Viruses would do better than their fellow cash machine viruses because they would be able to get around more efficiently. Bring on the music for the *Twilight Zone* . . .

What all this implies is that the non-arbitrary law-abiding sensibleness embodied in cash-machine button usage is being literally mapped, or reflected (there's that word again!), or translated, into the DNA carried by the Enter Button Virus, the DNA bearing the coding instructions for the virus's idiosyncratic behavior. Such a sly Enter Button Virus would, in effect, have become uncannily wise to the systems of sense and order inherent in its environment. It will have learned (albeit unconsciously), through its variant DNA, how to make good sense of specific non-arbitrary human activity. Recall that making sense is linked with adaptation and fitness. Thus the virus would be selected because it has adapted quite brilliantly to one particular sensible aspect of its environment. Of all the possible genetic variations in button preference, the genetic quota of this particular virus will embody the most sense.

The principle in operation here is the same one we met in chapter 3 where we learned about the fish eye algorithm. It was clear that the evolved virtual fish eye embodied the shape best able to reflect the sense, or order, in its virtual environment. The final evolved shape was the best way to fulfill the function inherent in the system in which it was embedded. With our hypothetical Enter Button Virus, the same principle is at work, namely the reflection and mapping of sense. In this case, the Enter Button Virus is selected precisely because it is making sense, or tuning in, to the sensible order prevalent in its environment.

As it happens, the same can unfortunately be said of today's AIDS virus, whose rapid evolution has been well documented. As sexual habits changed in order to curb its spread, so too did the virus, through variation, adapt itself to the less promiscuous behavior of its host, evolving in such a way that it now kills more slowly, thereby affording itself a better opportunity at transmission (as with the aforementioned "good luck virus," it is not out of the question that such viruses eventually evolve to become symbiotic rather than parasitic).

The eerie world of viruses shows that even the most primitive proto-life embodies some degree of intelligence, even if it is severely limited in its sense-making capabilities. Moreover, this intelligence is not static. Since variation is continuous and since certain aspects of the environment are always changing, this DNA-writ intelligence is likewise always changing. The learning never stops; natural intelligence is always active, always finding ways to become smarter, always feeding back on itself, always reflecting itself, always weaving itself in novel directions. And while a hypothetical Enter Button Virus might appear absolutely bizarre and untenable, it is actually *no more fantastic than real flu viruses and their ability to get themselves transported from host to host by way of coughs and sneezes.* The ostensibly outlandish principle remains exactly the same. A flu virus operating strategically at the back of the nasal passage—as opposed to elsewhere—is exactly like an Enter Button Virus adapting itself to a more commonly used button. Either example, the real virus or the imagined, is an equally remarkable instance of the sort of thing at which natural intelligence excels. Evolution is a learning process, or, as Margulis and Sagan put it, evolution is good at solving the problems of survival, the solutions being written in robust and heritable DNA. And the humble flu virus, a "simple" particle of life, or at least proto-life, still manages to carry a certain degree of intelligence within its store of DNA. That which lives is that which makes sense. If an organism cannot make sense, then it is out of luck and out of life.

BACTERIA

The Ultra-Smart Foundations of Life

Moving up a level in scale from the brutish and utterly selfish virus, we find bacteria, organisms more prolific than viruses (one spoonful of garden soil will likely contain more than ten billion bacteria!) but equally invisible to human perception. Bacteria are also redolent in natural intelligence. However, bacteria are considered a bit nasty by the human race—we talk of them as being "germs," unseemly entities that linger in nooks and crannies and that need to be remorselessly destroyed. Indeed, we lump them together with viruses. This is assuredly a case of bad public relations, since the role of bacteria in the general global health of the biosphere is crucial. Indeed, bacteria *invented* photosynthesis, assuredly one of the smartest processes within the tree of life. Furthermore, en masse, the wealth of photosynthetic bacteria, which evolved billions of years ago and still proliferate, helped to create an oxygen-rich atmosphere, which galvanized and energized the subsequent evolution of plants and animals.

Some species of bacteria are also able to *fix* nitrogen, that is, they can recycle otherwise "lost" atmospheric nitrogen into organic molecules that animals and plants use to manufacture protein (and DNA). The natural form of nitrogen is two atoms joined into a gaseous molecule. The kind of triple atomic bond holding two nitrogen atoms together is among the toughest in Nature, and only something like the massive power of lightning is able to break such a bond under non-life circumstances. Yet bacteria have learned to break this bond and thus help establish a life-confirming global nitrogen cycle without which the tree of life would perish. Unseen molecule-manipulating bacteria therefore sustain the most fundamental parts of the life's web. They are the most prodigious recyclers of the planet, their metabolic activity diverse and astounding. As Margulis and Sagan attest:

> Bacteria can swim like animals, photosynthesize like plants, and cause decay like fungi. One or another of these microbial geniuses can detect light, produce alcohol, waft hydrogen and fix nitrogen gas, ferment sugar to vinegar, convert sulfate ions or sulfur globules in salt water to hydrogen sulfide gas. They do all this and much more not because they are "pathogens" or in service to clean our human environment but because their survival imperative led to their inventing every major kind of metabolic transformation on the planet.[1]

This is the expressive flow of natural intelligence as it develops and progresses over the aeons, controlling the evolution of bacterial species, each member of which represents an autopoietic intelligent structure able to make sense of Nature and thus actively *do* things, as opposed to non-living matter that, by contrast, passively succumbs to physical and chemical forces. Further:

> Bacteria have already mastered nanotechnology; already miniaturized, they have control of specific molecules about which human engineers dream. Far more complex than any computer or robot, the common bacterium perceives and swims toward its food. Choosing and approaching its destinations, bacteria propel themselves by flagella, corkscrew-shaped spinning protein filaments attached to living motors in the membranes of their cells. Complete with rings, tiny bearings, and rotors, they are called "proton motors" and spin at about 15,000 rpm. These proton motors move bacteria in the same way that "electric fan" outboard motors propel boats.[2]

The tragedy is that we cannot perceive this sort of wonderful natural technology unless we happen to have access to a powerful microscope. Like so much of natural intelligence, the bacterial variety remained invisible and unattested for the entire course of human history until human intelligence "caught up" via the invention of the microscope in the sixteenth century. Moreover, the role of bacteria in recycling gases and controlling the atmosphere and greenhouse effect within the biospherical web of life has only become appreciated relatively recently. Never has there been there such a clear case of an invisible life support system at work all around us. Bacterial manifestations of intelligence saturate the globe and function as the very infrastructure of the biosphere.

THE ANT AND THE SUPERORGANISM

From invisible magic to visible magic. Enter ants. Everyday ants offer us more directly observable substantiations of natural intelligence, although even these may pass us by unless we are astute. Ants and the colonies of which they are a part are fantastic for any number of reasons, and it is rather unbefitting that hordes of them are forever being squashed beneath careless feet. Take their ability to lay down pheromone trails for instance (pheromones are chemical messages, scents laden with sense), a behavior that eventually leads to the formation of ant paths. This is a classic manifestation of *emergent* intelligence, a form of intelligent behavior that arises out of the interactions of lots of ants in concert. So not only do ants represent physical orchestrations of intelligence in terms of their individual biological structure, but the behavior orchestrated *between* them represents a further level of intelligence.

With trail-making, what happens is this: Ants leave their nest in order to go foraging. If you see ants whizzing along the kitchen floor or along the kitchen table in the summer months, this is what they are doing. Foraging. Shopping for discarded crumbs of cake and congealed drops of lemonade. If an ant locates a food source, it begins to collect the food and take it back to its nest. As it does so, the ant simultaneously exudes a special chemical from a gland inside its body, thereby laying down a chemical trail to and from the food source. If another ant walks across this chemical trail, it will automatically follow it and be led to the food source. Then it too will lay down the self-same chemical message as it returns food to the nest. And so on. Eventually an enduring chemical gradient will be established (the resolution of a chemical path) as more and more ants begin moving to and from the food source. This results in a sight that we must all have witnessed, namely long lines of ants ferrying food in an overtly professional manner. If this takes place in your home it might be a disturbing sight, a trail of relentlessly marching ants paying no heed whatsoever to the fact that they are invading a human home.

Although this kind of trail-making behavior is emergent, with each individual ant not necessarily "consciously knowing" what it is doing, the behavior is nonetheless indicative of the intelligence hardwired into ants. Somewhere in the DNA of ants lie instructions for building nervous systems and chemical systems that come to life when the right environmental cues are present. But we should bear in mind that the automatic trail-laying and trail-following behavior does not make sense in terms of single isolated ants; rather it makes sense only in the context of a colony of ants (and the environment) whose members continually interact. Such behavior can be thought of as a kind of distributed intelligence that emerges when certain conditions are encountered.

Unsurprisingly, some ant specialists like Harvard professor of entomology E. O. Wilson have coined the term "superorganism" to account for the collective social intelligence of an ant colony. Termite and bee colonies are likewise superorganismic in nature, the interactive behavior of individuals allowing the colonies they comprise to survive. Consider the temperature regulation of a bee colony. By bees collectively fanning their wings more or less, a beehive can control its internal temperature. An individual bee does not need to consciously "know" that it is helping to regulate the temperature, yet hundreds of bees influencing one another by their individual fanning actions yields a collective homeostatic solution to the problem of temperature control.

CITY ANTS

The smart trail-laying behavior of ants can be observed by anyone with a garden and a little patience. As an intriguing instance of emergent intelligence built upon the hardwired behavior of interacting individual ants, it is instructive to watch the process manifest over time. I happen to be writing this book in a particularly busy part of London, but there is still a garden outside, a mini-section of wilderness in which insects thrive. One morning, in order to test the trail-laying capacity of London ants, I placed in the garden a small heap of brown sugar. This would undoubtedly appear to the ants to be some divine manna from heaven, an excessively rich source of food appearing as if by magic.

Having set up the bait, I stood aside and waited. Actually I had to wait quite a bit, so much so that a chair was called for. After ten minutes or so, nothing much had happened. I became a tad disillusioned. Perhaps, I mused, these London ants were devoid of the collective intelligence shared by their tropical cousins. Maybe London's pollution had gotten to them. Or maybe they were sickly inbreeders, a decadent strain no longer capable of trail-laying.

Of course, I had no need to fret. It went without question that these London ants could lay chemical trails. The fact that they existed at all decreed that they could perform in their usual collectively intelligent manner. It was only a matter of time. After about fifteen minutes, a few pioneers had located the manna and were avidly ferrying sugar crystals back to their nest. Soon, more and more ants were on the trail until, after about half an hour, the trail had become firmly established, had become, in fact, a kind of unmistakable ant thoroughfare. From sketchy beginnings, the trail had become more and more pronounced, straighter and straighter. The simple experiment was a success. An emergent form of natural intelligence had materialized. If the chemicals underlying the individual trails had been visible to human perception, they might have appeared like wispy bands of color slowly converging into tighter and tighter lines until eventually a straight line of concentrated color connected nest to food source.

A few days after my experiment I even observed trail-making behavior at an aboveground train station. In the midst of concrete, giant trampling passengers, and copious train fumes, a plucky colony of ants—presumably living under the platform somewhere—had located a discarded can of Coke. The trail led directly into the open top of the can. Oblivious to commuters, there must have been fifty or so ants moving straight across the platform from some crack to the pool of sugar-rich liquid in the can. Not exactly a glamorous spectacle of natural history, but a fine example of the tenacity of natural intelligence nonetheless.

Trail-laying chemicals are produced in ants by an organ called *Dufour's gland*. If one "milks" such a gland and lays down a "fake" trail, ants will be deceived and follow it. However, such deception will not reign long since there will be no reinforcement of the trail, since no food is found at the end of it. Also, trail-making chemicals dissipate over time (this represents a further smart aspect of ant-embodied intelligence, as the indefinite lingering of trail chemicals would lead to wasted journeys). The following of fake trails shows us that trail-making behavior is autonomous—it happens quite naturally and spontaneously among ants due to individual hard-wired behavior, the hardwiring having evolved according to the logical sense-making dictates of natural intelligence over millions of years.

Unsurprisingly, the AI scientists we met in the previous chapter have already written papers on substantiating this kind of emergent collective intelligence in their robotic creations. Some have speculated upon the idea of landing a colony of small (and thereby relatively cheap) mining robots on a planet. The little robots scurry off in all directions, and if one locates some suitable ore, then it returns to the mother ship and leaves some sort of trail in its wake. Other robots in its vicinity who steer across such a trail then automatically follow the trail to the source of ore. Then they return with the booty and lay down a similar trail. The general idea is exactly the same as with ants—intelligent behavior is orchestrated from non-random individual actions.

Other glands in ants produce poisons, alarm substances, and grooming-inducing chemicals. This is how ants communicate—through a language of chemicals, the language, like any language, making sense only in the context of other ants. There is even an ant chemical whose linguistic meaning is concerned with funeral rites. If an ant dies inside a nest, then the fatty acids produced by the decomposition process act as a message that other ants cannot fail to ignore. If they sense this chemical signal—the cry of "I am dead!" issuing from the deceased—then a response is elicited in which the ex-ant will be picked up and deposited outside of the nest. Experiments have been carried out in which live ants have been coated with this molecular death signal and then been reintroduced into their colony. Although they are still alive, such unfortunate ants are nonetheless seized and ungraciously dumped outside the nest, repeatedly so if they attempt reentry. Although this might suggest that, in this instance, ant behavior is a bit stupid, the scenario would never transpire under natural conditions. What it does highlight is the sheer robustness of ant intelligence, that successful strategies for sensible living have been scripted in ant DNA. That ants exist throughout the world and have done so for millions of years testifies to the success of ant-embodied intelligence.

Other ingenious aspects of ant behavior come to mind. If you are ever out amid unspoiled countryside, it occasionally pays to overturn small rocks. For underneath such rocks one often comes across a medley of insectile intelligence. Ant colonies are often found there. If you find such a nest, then within minutes any exposed eggs on the surface of the nest will be taken belowground by workers in a flurry of sensible behavior, even though the removal of their "cover" is an event unlikely to transpire under normal conditions. But the main item of interest here is that the establishment of ant nests under rocks is no accident but another mark of natural intelligence. Rocks tend to absorb early morning sunshine. They heat up. Ants are innervated by warmth—that's why they come out in force in the early part of the summer as soon as it gets warm, often invading our kitchens out of the blue. An ant nest located under a small rock will heat up relatively earlier than a nest positioned elsewhere. Since an ant colony under such a rock will be roused earlier and will be the first to search for food, they will have first pickings of any available food source. Hence such a rock-dwelling colony will be selected by Nature. So this is the wholly logical reason for finding ant colonies under small rocks. Once again, we can see that ants have made great sense of the contextual environment in which they exist. Their hardwired tendency to select small rocks under which to build a colony *reflects* the sensibleness of the natural world, the fact that small rocks heat up under the morning sun.

THE ULTIMATE IN ANT INTELLIGENCE

Perhaps the most extravagant manifestation of intelligence found in the ant world is that embodied in a species belonging to the leaf-cutter genus *Atta* (we have E. O. Wilson and B. Hölldobler's fascinating book *Journey to the Ants* to thank for the following information). The species in question actually cultivates a certain mushroom—in so-called fungus gardens within their nest—upon which they feast. They have acquired their leaf-cutter name from the fact that they collect and cut up leaves in order to provide mulch for the fungus. Colonies can reach millions in size, consuming as much vegetation as a cow. Alone, these are impressive observations. However, the life cycle of these ants demonstrates an even more remarkable example of natural intelligence, mathematical in its precision and faultless in its logic.

When reproducing (the process wherein virgin queens and sperm-touting males take to flight), a queen who gets inseminated stores a couple hundred million sperm, which can be kept upon her person (for up to fourteen years) and dispensed when necessary to produce offspring (i.e., eggs). Such a potentially fecund queen needs to burrow into the ground in order to establish a new colony. Before she digs, she tears off her wings, which would otherwise get in the way. She then digs a hole and, later, a small chamber in which to grow the fungus with which her species associate. This is possible because before she left her old nest she had actually taken a small wad of fungal hyphae (hyphae are the underground filaments of a fungus) and kept it in her mouth. In her newly dug chamber she spits out this important package of fungal hyphae so as to plant it. Then she lays her first eggs. So far so good; a new colony is in the making. But a serious problem is looming, namely the threat of starvation to the queen. At this moment in time she has to subsist solely upon the breakdown of her wing muscles and her fat stores, all the while laying eggs. She can last no more than fifty days on such meager nourishment as her recycled wing muscles and fat stores afford. And yet within this fifty-day period a successful queen will manage to raise workers who can go out and forage, feed her, and also help cultivate the fungus garden. All being well, eventually a new colony will be in full swing, consisting of different workers, each with a different function. Big foragers will be responsible for slicing leaves into bits; smaller workers will cut up these bits of leaf into smaller fragments; even smaller workers will tend to the fungus garden by molding leaf fragments into moist pellets; and soldiers will serve to protect the colony from attack.

The most incredible aspect of this royal drama surely lies in the collection of fungal hyphae by the queen before she leaves her old nest on her maiden flight. She does not have to consciously "know" what she is doing, of course (although ants might enjoy some sort of cognitive experience). It is not that she says to herself, "Ah, methinks I should take some fungus samples with which to instigate a new fungus garden in my planned new nest." Her smart behavior is presumably unconscious and autonomous, just as autonomous as her "desire" to take to the air, get inseminated, locate a suitable spot to start a new colony, and dig a new burrow. It all comes quite naturally. Yet all this eminently sensible hardwired behavior—especially the collection of fungal hyphae for future transplant—is utterly extraordinary in terms of insectile savoir faire. Indeed, one could say that the ant's hardwired behavior is a manifestation of "understanding"—that "understanding" does not have to be conscious but is, rather, one aspect of the ant's intelligence (just like the wing structure and wing musculature of a bird represent an "understanding" of aerodynamics). The organization of countless atoms and molecules into the macroscopic autopoietic conglomeration that is the queen ant and all her sensible behavior is a classic example of organic intelligence—even though it may not be described as such by conventional science.

We might get to wondering about how a leaf-cutter ant colony can produce the different types of workers mentioned. How is this capability determined? Since different types of workers will be required at different stages of a colony's development, a full complement of workers becomes necessary only when a colony is mature. Thus, there must be some clever control mechanism at work. As it happens, E. O. Wilson did experiments to explore this issue. Through "trimming" the size of a leaf-cutter colony, Wilson was able to fool it into thinking that it was much younger than it really was. The colony subsequently became "reborn," as it were, rejuvenated, producing only the sort of worker needed in a young and immature colony. This experiment shows that by some contextual means the size of a leaf-cutter colony feeds back on itself and determines what sort of worker the queen produces. This is more evidence for emergent intelligence wherein a system of interacting ants is self-sensitive (i.e., cybernetic) such that intelligent behavior manifests, on many levels. Not only is an ant colony a cybernetic superorganism, it is akin to a natural software program, or computation, which is tangible, and which flows in space and time, a colony of ants representing a kind of fluidic pool of self-sustaining intelligence. When one learns about the life and behavior of seemingly simple ants, locating a colony out in some wild wood becomes a real find.

THE ART OF HACKING INTO THE ANT SUPERORGANISM

Even the predators of ants pay testimony to the cunning manner in which natural intelligence weaves and reflects itself. The assassin bug (*Ptilocerus ochraceus*) is so named because of the passive-aggressive tactics it metes out to the Asian ants upon which it feeds. Assassin bugs simply sit in or near an ant path, where they gently waft an intoxicating attractant from glands on the underside of their abdomens. This is not an act of innocence but chemical subterfuge. When a worker ant approaches—doubtless on its way to some food source and diligently following a chemical trail—the assassin bug raises itself upon its hind legs and presents the surface of it alluring aromatic gland for closer inspection. Entranced, the worker ant is lured away from its trail and moves over to examine the bug's gland. Unable to contain itself, the ant begins to lick the gland's chemical secretions. The assassin bug then carefully folds its front legs around the body of the ant and brings the tip of its daggerlike beak into position at the back of the ant's neck. The ant is oblivious to the impending fatal embrace and continues to lick at the bug's intoxicating gland.

The ice-cool assassin bug does not kill the ant right away but remains tentatively poised to do so. After a few minutes of feeding at the gland, the ant begins to show signs of paralysis, perhaps alike in nature to some drunken stupor. When completely helpless, having curled up into a ball and drawn in its legs, the ant is quickly killed by the bug. The ant's neck is pierced and its life sucked out. When the meal is finished the assassin bug does not retreat but stays in the vicinity in order to procure more nutritious victims, the marching trail of ants oblivious to the daylight "roadside" murders taking place in their midst.

Given the chemical language used by ants, the assassin bug's intoxicating secretions, so attractive to the ants upon which it feeds (reminiscent of the anal secretions of aphids of which ants are fond), are obviously linguistic in nature. Just as the ant victims are picked off while traveling along a chemical trail akin to a sign-posted path declaring THIS WAY TO FOOD, FELLOW ANTS!, so too are the chemicals emanating from the assassin bug's glandular weapon akin to a language, moreover a language that ants are deceived by. Perhaps the chemical message is interpreted by the victims as NICER FOOD RIGHT HERE, LADS! Whatever the case, the message must be alluring and in the right "dialect"; otherwise an ant victim would not be fooled. The chemical message has to be very similar in nature to the messages normally encountered by ants.

In other words, what the assassin bug is doing is *hacking* into the ant superorganism. If we consider an ant colony to be like a giant distributed nervous system in which information is processed chemically (and E. O. Wilson has drawn such analogies), then the assassin bug is hacking into the ant system for its own parasitic ends, just as a human hacker can hack into some database and steal funds. Which means that part of the assassin bug's DNA-writ genome is concerned quite literally with hacking, of the chemical kind, into the similarly DNA-writ collective genome of an ant colony. Put in these terms, natural intelligence once more comes to mind as the most suitable paradigm for accounting for this kind of remarkable biological data. To merely shrug and bandy the word "evolution" in the air does not do full justice to the canny insectile wile that we here observe.

MAGNIFICENT MOTHS

The degree of natural intelligence substantiated within ants and their colonies is matched by that of other insects like moths and bees. For example, consider both the design and the sensitivity of the antennae of the male silk moth *Bombyx mori*. It has been established that each of these antennae are covered in some seventeen thousand odor-receptor hairs called sensillae, which are arranged in a fractal branching manner much like the way bird feathers are constructed. Each hair, or sensillum, has about three thousand pores, each of which can capture *individual* airborne molecules. Therefore, a male silk moth is equipped with a total array of 102 million molecule-sensitive pores, a biological form of nanotechnology stunningly advanced in nature. If enough of the "right" molecules are caught—usually deriving from a female silk moth—then a nerve impulse will be generated, which will cause subsequent chains of nervous activity eventually resulting in a definite pattern of behavior in the male.

These hair structures are actually so sensitive that they can detect a single pheromone molecule released by a female silk moth one mile distant. This outrageously sophisticated form of communication is worth repeating in another way. A sexually charged female silk moth—out to secure the charms of a male—releases pheromones, linguistic chemicals similar to those employed by ants. By utilizing the natural process of diffusion—random thermal motion—this chemical message will spread through the air, a kind of deliberate non-random exploitation of otherwise random physical processes on the part of the moth. After a certain amount of time has elapsed, molecules of this message will have been broadcast for miles around, a pretty neat trick when you think about it, since it is a highly energy-efficient means of communication. Just think how much more complicated and tiresome it would be to set up a PA system or to shout ever so loudly over and over again. At any rate, if just a hundred of these molecules reach the antennae of a male silk moth and become registered, they will elicit a nerve impulse that will cause a growing cascade of nerve impulses to fire—the sensible amplification of the received signal, in other words. Thus, natural intelligence is here shining at both ends of the moth interaction. Once the message has been fully assimilated, the male moth will start actively searching for the female by making use of the non-random chemical gradient associated with the female's signal, just as the ants previously discussed utilized chemical gradients in order to hone in on some location.

The noted nineteenth-century French naturalist Jean-Henri Fabre found all this moth magic inconceivable, declaring, "One might as well expect to tint a lake with a drop of carmine." We can, perhaps, forgive Fabre for his denial of natural intelligence. After all, a mile is a relatively long distance to a moth. How could a male moth possibly detect the presence of a female from such a vast distance without recourse to a radio transmitter or a mobile phone? It does seem improbable. Yet natural intelligence continually specializes in this sort of exquisite biological engineering wherein the sensible properties of physics, chemistry, and natural law are made sense of through biological technology. Indeed, such magnificent moth-brand intelligence goes on all the time regardless of the fact that, for us, it is mostly invisible. When we see a moth on a tree or on the kitchen wall in the late summer months, we are generally oblivious to the fact that an intimate sexual molecular language may be unfolding between moths up to a mile or so apart. Considered in toto, the airborne chemical messages of all insects (and there are lots of insects, some fourteen billion per square mile!) must exist alongside one another in the same way that our multifarious radio signals exist alongside one another. Refined telecommunications technology is far more ancient than we realize.

BRILLIANT BEES

Language is a hallmark of intelligence, for language involves the communication of information. And it is information that is the key currency being manipulated in all transactions that are construed as being intelligent transactions. DNA is a language, as is human speech. The pheromones released by moths and ants are examples of a language. Informational elements within a language, whether these elements are words, genes, or specific molecules coded by genes, convey some message, some signal, some kind of meaning. They *do* something; they cause significant things to happen. As far as I know, colonies or tribes of tangible robots have not yet been developed that can communicate with one another through some sort of overt language. In theory, they could use radio signals or flash neon colors at one another to convey information. To use language, to affect the behavior of another individual without directly touching that individual, represents a quite complex ability, an action at a distance or an extended phenotypical effect as Richard Dawkins has called it (a phenotype being the full expression of an organism's genotype).

We have already seen how a chemical language is utilized by insects like ants and moths, but bees have gone a step further and make use of an elaborate tactile and auditory language. Information transmission according to touch and sound, in other words. This rather extraordinary manifestation of natural intelligence may well have gone unnoticed by modern humans for the first hundred thousand or so years of our existence, despite our penchant for honey. It was only in the past century that naturalist Karl von Frisch discovered that bees "dance" in order to let their nest mates know where a food source is located (for this work, Frisch received a Nobel Prize, whereas Nature itself, as usual, got no reward). Or, to put it another way, we could say that this particular aspect of natural intelligence managed to *introduce* itself to Frisch. In any case, the dancing behavior of bees is actually a bona fide language, since it has been firmly established that such dances convey information regarding the specific whereabouts of food. In a nutshell, a bee who has located a prime food source like nectar will return to its hive and then commence dancing. Its nest mates will flock around, all the while sensing and listening to the dancer's movements. The location of the food is conveyed in the movements and vibrations of the dancer.

Different species of bees use different varieties of dance. Frisch likened these varieties to different dialects in bee language. Some so-called waggle dances are basic in that the dance is done on a horizontal surface, whereon the dancer bee faces the direction of the food source, which was noted according to the position of the sun on the bee's initial outward journey from the hive. Its nest mates can then fly off in that direction. Other dances are performed on a vertical surface. In this case, information pertaining to visual direction is *transposed* onto gravitational coordinates. Which is to say that if the food source is directly in the direction of the sun, then the dance is done absolutely vertically, all the movements pointing straight upward. A food source at, say, 45 degrees west of the sun would be indicated by movements at 45 degrees to the left of vertical, and so on. As to the distance to the food source, this is conveyed in the *speed* at which the messenger bee dances. *Even the constant change in the sun's position is taken into account by the dancer.* It transpires that a dancing bee will adjust its dance as time goes on so as to compensate for the sun's changing position. Finally, we should note that the entire dance will take place in the *darkness* of the hive.

How is one to respond to this kind of remarkable phenomenon? Our first inclination might be to deny it all, just as Fabre denied the ultra-sensitivity of male silk moths. Frisch simply had to have made the whole thing up, perhaps as an imaginative ruse to attract academic attention and some lucrative funding. For are we to believe that bees essentially "chat" to one another about the precise grid-referenced whereabouts of nectar-rich blooms? Are we to further entertain the even more fantastic notion that they can transpose directional information into a system of gestures, just as we humans can transpose directional information into sign language? Once again, natural intelligence is the best paradigm with which to account for such data, the dancing language of bees representing a collectively realized system of active intelligence in which sense is made, recorded, and transmitted.

The more one dwells upon insects like bees, the more natural intelligence becomes apparent. For instance, how does some avid harbinger honeybee find its way back to its hive after having located a prime food source? Before it can convey its good tidings, it must first make it back to its hive, which might be all of two hundred meters away. It turns out that bees (and ants for that matter) can navigate by making use of *polarized* light, part of the light spectrum that we cannot perceive with the unaided eye. The compound eyes of bees carry out a kind of geometric analysis of polarized light from the sun, which is visible even if the sky is cloudy. This means a data record is kept by means of which a bee can "know" its position relative to its hive regardless of overcast weather conditions. Through some sort of clever mathematical operation, a bee can therefore retrace its route by making use of the potential geometric information in polarized light. How this kind of calculation is readily computed by a bee's tiny brain, along with numerous other intelligent calculations, remains an awesome mystery. It would take a genius and a truly massive textbook to fully detail the quota of natural intelligence in operation within a "humble" bumblebee.

As for the high-octane fuel used by honeybees, which enables them to fly every which way, it has been estimated that if they are running on honey—the end product of the processing of collected nectar—then they can fly two million miles to the gallon! In other words, a speck of honey on one's finger is enough—in the context of a bee's metabolic system—to fuel it on a flight of one thousand miles, and with no ensuing asthma-inducing exhaust fumes either. This is naturally intelligent eco-engineering at its non-polluting best.

Returning once more to the impressive dancing language of bees, the most primitive form of dancing encountered by Frisch was that found in a tiny species of stingless bee. Frisch discovered that once a food source had been located by this species, the bee with the information would return to the hive and knock against its hive mates, but only so as to arouse them. No information about the distance or direction of the food source was conveyed, only the lingering scent of the flower and the stimulus to go out and forage. The rest was up to chance, since the bees so stimulated would have only a trace of flower scent to guide them.

If we assume that this basic form of "dancing" is indicative of the very evolutionary beginnings of refined bee language, we can appreciate that this is just the sort of behavioral system through which natural intelligence can become further concentrated and further elaborated. Since there is a wealth of potential sense to be made by a beehive in constant need of nourishment, the ongoing play of natural intelligence will ensure that more and more of this survival-dependent sense will become reflected or mirrored by the bee superorganism, just as the primate brain has become more and more adept at sense-making over the last few million years of its evolution. Honed over vast spans of time, a basic bee language like that of the tiny stingless bee will eventually attain the sort of linguistic competency we witness in other bee species. Natural intelligence pushes out in all directions, and any biological direction with the potential for further elaboration in terms of sensemaking will always be investigated and ramified, the result being the consistent realization of ultra-smart systems of biological behavior. Left for a few more millions of years of evolution, bees might even eventually be able to discuss strategies for avoiding eco-destructive humans.

TERRIFIC TERMITES

If you happen to be a termite living in a termite mound, the control of temperature and humidity are crucial considerations. An entire African colony will perish if exposed to dry air. Therefore, such colonies must forever ensure that appropriate levels of temperature and humidity are maintained within their mounds. It has already been mentioned that bees control the temperature of their hives by means of fanning their wings, being so adept at this that they are quite literally a pandemic species. However, termites, by virtue of their different morphology (termite workers have no wings), are forced to use a different technique, albeit one equally as reflective of natural intelligence.

One method of humidity control is to exploit the physiochemical properties of water. By making use (i.e., making sense) of the evaporative property of water, cooling can be achieved (that's why we and many other creatures sweat). Termite colonies have learned to incorporate this method of air-conditioning inside their complex mounds. As a strategy it is noteworthy as it once again shows how the sensibleness of physical and chemical law—here expressed in the specific behavior of water molecules—has become reflected, or translated, through equally sensible biological systems.

In his book *Alien Empire,* Christopher O'Toole describes the most salient aspects of termite mound architecture, which include the use of water evaporation as well as molecular diffusion. The termite species in question also happens to associate with a fungus, which it cultivates and dines upon:

> The mounds of some species are huge—7.5m (25ft) high in the case of *Macrotermes bellicosus*, in Africa. The earth mound is built above the nest chambers and fungus comb. Hot air from these living quarters rises and flows up a central chimney and from there, via side branches, to a system of thin-walled tunnels in the outer wall. Here, where there are only a few millimeters of outer wall, the air cools, carbon dioxide diffuses outward, and fresh air diffuses inward. The continuous flow is maintained by pressure from the hotter air rising from the nest proper.
>
> The flow of air now enters a large chamber beneath the surface of the soil, passing between a series of large vanes, which the termites keep damp. This dampness cools by evaporation, and the fresh air passes back into the living area.
>
> It is remarkable that the worker termites have constructed the equivalent of, in human terms, a skyscraper 9.6km (6 miles) high. And they are blind.
>
> The air-conditioning systems of termites are so effective that human engineers are now constructing buildings with cooling systems based upon termite design.[3]

There are even termites in the parched deserts of northern Australia that build mounds according to the overhead orientation of the sun. In order to minimize the sun's sweltering heat during the day, this particular species (aptly called the compass termite) constructs a wedge-shaped nest, which is aligned north and south. When the sun is directly overhead during the day, only the narrow ridge at the top of the nest is exposed, whereas during the morning and the evening the large flat sides of the nest absorb the full heat of the sun. The result is an uncannily effective architectural method of temperature control, the sensible and precisely orientated architecture matching the sensible course of the sun. Similarly, in the rain forests of Gambia species of termites construct mushroom-shaped nests, this shape allowing the copious rain that falls in the rain forest to flow off the nest without causing damage. Both types of nests admirably exemplify the sense-making building behavior of termite colonies, the particular sensible designs matching the particular sensibleness of the environment in which the nests are built.

In the former case—the wedge-shaped nest aligned north to south—this is an evolutionary adaptation made possible only given the non-arbitrary motion of the sun. Which means that the lawful course of the sun relative to the Earth—determined as it is by the law of gravity—represents one particular facet of the sensible context repeatedly referred to in previous chapters. Because the laws of Nature are not haphazard, they ensure that sense can be made of their effects. In the case of gravity and the myriad sensible things gravity does to gross matter (like determining precise planetary revolutions about the sun), the equally sensible responses of bio-logic to gravity are numerous. Not only do we find precisely aligned termite nests, we also find bone structures, gaits, cellular processes, and all manner of biological processes mirroring the intelligibility of gravity.

The key principle to hold in mind is once again that of an intelligently configured contextual system (i.e., all of Nature) fostering the evolutionary emergence of biological systems *wise* to that system. The same goes for the use of water evaporation for cooling. The precisely dependable behavior of water molecules is a sensible reflection of sensible physical and chemical laws. Biological systems able to make survival use of the evaporative properties of water are therefore mirroring, or reflecting, the intelligibility of chemical law. Natural intelligence would thus appear to saturate everything, reflecting itself, mirroring itself, and reverberating through systems as diverse as physics, chemistry, and biology. Life, in all its manifest glory—whether evinced by termites, ants, bees, moths, or bacteria—is clearly replete with intelligence, biological evolution being an auto-response of Nature to its own inherently smart configuration. The ultimate conclusion is worth reiterating: Nature, in its entirety, is a system of self-organizing intelligence.

BOTANICAL INTELLIGENCE

Let us not forget the natural intelligence of plants. For many people the most striking plants in the world are those able to catch and ingest insects, namely carnivorous plants. No one can fail to be impressed by this kind of unusually crafty botanical behavior. For while we might tend to think of vegetation as passive, static, something we eat, stomp on, brush aside, and so on, carnivorous plants force us, willy-nilly, into admiration. Indeed, their carnivory is introduced here as a particularly fine example of natural intelligence more readily acknowledgeable than many of the other adept properties of plants like their ability to seek light and water.

There are hundreds of species of carnivorous plants dotted about the Earth. It is interesting to learn how their carnivorous appetites first became known. It appears that when botanists first studied and classified these plants, they failed to realize their carnivorous nature. And these botanists did not just fail slightly in their documentation of this particular manifestation of natural intelligence—they failed miserably. Which is to say that the trapping and ingestion of insects by such plants went completely unnoticed when they were first named and classified. Late eighteenth- and nineteenth-century botanists, while being adept in the booming practice of plant classification (taxonomy), were nonetheless poor in their appreciation of the scope and scale of the natural intelligence embodied in their study material.

Take sundews, for instance (a species of *Drosera*). Small, squat, and ruddy, these amazing plants grow in wild and wet ground all over Europe. Their tiny leaves have lots of reddish hairs, which exude a sticky substance (the dew) that will trap any small insects landing on their surface. This species was first described in 1780, which means it was officially named and its most salient properties noted. Or so it was thought. For the plant was not fully understood till 1875. In other words, it took a hundred years for botanists to realize the naturally intelligent function of the sundew's hairs—that these blatant sticky protrusions are able to glue insects and then actually *bend around* the trapped meal so as to better digest it, this digestion taking place on account of the digestive secretion of the hairs (the hairs are thus equipped to secrete two functional substances).

Actually, it was the ever patient and ever perceptive Darwin himself who first documented this remarkable expression of botanical intelligence. One must assume that before Darwin's time, botanists were incapable of conceiving that certain plants had the ability to trap and digest insects. Such a concept would have been deemed improper and more in line with fantasy than scientific truth. How could a mere piece of immobile vegetation secure and thence eat prey? One hundred years of ignorance. And don't forget, this was a post-Newton period. It is not like late eighteenth-century science was immature. All sorts of breakthroughs and insights were being made into the fundamental secrets of Nature. Yet the shrewd organic intelligence embodied in the humble sundew remained elusive.

What about pitcher plants? One cannot fail to be awed by tropical species of *Nepenthes,* adorned as they are with large ornate "jugs" whose function is now known to be the trapping and consumption of insects, which drown within the pitcher's water-laden interior. The year 1737 marked the time when botanists first noted the insects found floating in the pools of water within the pitchers of these plants. But it was not until 1791 that it was first cautiously hypothesized that some sort of digestion might be taking place. Indeed, it was not until 1829 that science suggested that pitcher plants could digest insects in the same manner that the mammalian stomach digests food. Almost a hundred years again; a century for human intelligence to fully grasp and acknowledge this particular manifestation of natural intelligence. Certainly Linnaeus—the esteemed father of botany—would have none of it. He simply refused to believe that plants could dine upon arthropodic meat.

His mind could not stretch to incorporate such heretical notions.

The same holds true for the Venus flytrap, perhaps the most distinguished of all carnivorous plants. In 1769 the botanist J. Ellis wrote to Linnaeus about this species, suggesting that when insects stopped struggling from the grip of the plant's flytrap, then the leaf lobes opened and let them be on their way. Sounds somewhat silly, but at least it stops us from having to accept that the Venus flytrap eats insects like some animals do. Similar incredulity followed a century later in the 1870s when it was shown that the electrochemical impulses generated in the Venus flytrap could be triggered by hair cells on the plant's trap mechanism, impulses of the very same kind as the nerve impulses transmitted within the human body. It seemed bizarre and unseemly. Thankfully, despite murmurs of disbelief, science soon recognized the true function of the flytrap.

It is clear then that the botanical products of evolution are often far smarter than we imagine. Indeed, if we go back to the pitcher plant, modern research has shown that special cells in the interior wall of the pitcher provide extra oxygen in order to keep the pitcher's store of water fresh and suitable for sustaining the life-forms that live in the water and initially process the drowned insects. Once the insects are predigested in this manner, pitcher plants can then absorb the processed insectile residue to their heart's content. It is a very clever system and not deserving of the term "designoid," which Richard Dawkins explicitly uses to account for the botanical wizardry actualized in pitchers.

Another example of the ability of carnivorous plants to outsmart human observers is evinced by bladderwort, a simple-looking pond plant that garners insects on which to feast by way of cunningly designed underwater traps. Although minuscule, these traps are, like the pitcher of a pitcher plant, strategically enfolded leaves. In this case, though, the leaves are structured in such a way as to form small water-filled sacs with a one-way entry, which will entice and subsequently trap any curious insects who chance upon them. Botanists now describe the sophisticated trap structure with terms like "hinge," "door," "wall structure," "buttress," "bellows," "door tension," "tripping apparatus," and so on. All the parts of this sac structure are designed explicitly to trap insects. Afterward, the insects are eaten. In 1797 the eminent botanist J. Sowerby thought these tiny plant structures to be flotation devices of some sort. He even thought that the insects he observed inside the sacs had taken up lodgings! Only in 1875 was it discovered that the sacs were quite literally insect traps, able in some species to ensnare and fully digest prey within forty-eight hours, exactly as if they were miniature stomachs. And it has now been firmly established that the traps actually make use of a *partial vacuum,* which acts to suck victims inside. The notion of a plant able, through evolutionary honing, to mold its leaves such that an exploitable partial vacuum can be generated is enough to make one giddy with astonishment.

NATURALLY INTELLIGENT ORCHIDS

Notwithstanding the extremely refined natural engineering embodied by carnivorous plants, mention must here be made of another memorable example of almost unacceptably astute plant design. The example concerns the pollination mechanism of the bucket orchid. All orchids are extraordinary. Their pollen-emitting flowers often mimic female bees, being so accurate that male bees will try to mate with the flowers and inadvertently be covered in pollen. If a large orchid could fool *a man* in the same way, it would be newsworthy for years . . .

The bucket orchid has not merely perfected the sly art of sexual imitation but seems to actually be *showing off* its embodied legacy of intelligence. As naturalist Sir David Attenborough carefully explains:

> The bucket orchid grows high in the canopy of forests of central America, sitting on the boughs of the great trees. Its flowers are yellow, sometimes plain and sometimes mottled with orange or brown

according to the species, and they hang down, half a dozen or so on a stem. The front of the flower is formed by two small wings. These serve as signposts. Behind them hangs the little bucket that gives the plant its name. When a flower opens, two small glands on the stem . . . secrete a liquid that drips down and fills the bucket to a depth of about a quarter of an inch. The flowers now give off a sweet heady perfume. Each of the twenty or so species of bucket orchid has its own brand of scent. Although human nostrils cannot distinguish between them, little iridescent bees that live in these forests certainly can. Each species of orchid attracts its own species of bee.

It is only the male bees who respond to the orchid. Its smell seems to excite them greatly and when a flower opens there will soon be several male bees buzzing around it in an agitated way. Before long, one will land on the side or the rim of the bucket and make his way to a rounded pad that rises from the rim at the base of the short stem connecting the bucket to the front of the flower. From this pad he scrapes an oily substance, which he packs into pockets on his back legs. This is not food. It is an ointment that he will use to attract females during his elaborate courtship rituals—which is why each species of bee needs its own special brand.

Once he has collected a full load, he tries to fly off. But there can be such a congestion of excited bees in and around the flower, and the surface of the pad is so slippery, that sooner or later, one of the bees loses his footing and tumbles into the fluid at the bottom of the bucket. There is only one way out—a tunnel leading up through the front wall of the bucket to the outside world. On the wall just below this escape hatch there is a little bump. It is the only reasonable foothold to be found on the otherwise smooth slippery walls of the bucket and eventually the bee discovers it. Using it as a step, he climbs up and enters the tunnel. It is a very tight fit but he manages to squeeze his way up it. Just as he is about to emerge from the other end and regain his freedom, his back catches on a projection in the roof of the tunnel. This is caused by two lumps of the orchid's coagulated pollen called the pollinia. The bee continues to struggle forward and finally scrapes off the pollinia from the roof of the tunnel so that when he does at last emerge, they are attached securely to his back like a little knapsack.

Meanwhile, other bees are still feeding on the pad. One may have gone through this obstacle course before in another flower and already have pollinia on his back. If such a one slips into the bucket, then the orchid is lucky, for when he struggles up through the tunnel, a hook on the roof neatly engages on the pollinia and removes them. The orchid, after all the travails of the bees, has been fertilized.[4]

Can this scenario really be true? It was hard enough accepting the natural intelligence of bees. Now we must concede that bees are themselves coerced into further systems of intelligent contrivance, being politely but firmly encouraged to follow a kind of orchid assault course in which, willy-nilly, they help pollinate the orchid. Pollinia knapsack indeed! This is natural intelligence in its most wonderfully wild and organically ornate guise.

All in all, these examples of plant design—particularly those of carnivorous plants—reveal how much of natural intelligence remains obscure due to the inability of human intelligence to grasp the extent to which evolution through natural selection—the manifest methodology of natural intelligence—has honed organisms. Who knows what other aspects of natural intelligence have yet to make themselves known? Indeed, as we shall see at the end of the following chapter, the planetwide action of plants provides new and emergent forms of intelligence, which may well control things like the quality of the atmosphere, just as the en masse actions of ants, bees, and termites lead to emergent systems of homeostatic control.

KEW GARDENS
A Case of Paradigmatic Mismanagement?

In speaking of the unsung intelligence of plants, I cannot fail to mention Kew Gardens, the spacious botanical reserve in London. Kew is home to a multitude of plant species, some tropical varieties being housed in distinctly lofty Victorian greenhouses. I once did voluntary work at Kew back in the mid-1990s, at a time when its director was Sir Ghillean Prance. I learned that he once wrote a scientific paper delineating the pollination mechanisms of the giant Amazonian water lily. The large flowers of this water lily manage to capture and ensnare a certain species of beetle, which is attracted by the flower's nectar. After having attracted a beetle during the day, the flower closes at night, thereby trapping the beetle. During its incarceration, the beetle becomes covered in pollen (as well as getting its fill of nutritious nectar). The next morning the flower opens and the beetle flies off, lands on another lily flower, and, being covered in pollen, inadvertently pollinates the flower.

This is a typically fine example of botanical intelligence, so stunning in its logical deliberation that Professor Prance was applauded for being the first to detail the pollination process in full. One can even view the giant Amazonian water lily as having *taught* the young professor. And yet the beautiful botanical establishment over which he once presided is devoid of clear-cut allusions to natural intelligence, the very process that makes the gardens possible at all, as well, of course, as being the impetus behind the academic career of most of the staff there.

Kew boasts the Evolution House, a large indoor exhibit documenting the process of evolution. Yet a stroll through Kew's Evolution House reveals no explicit mention of intelligence. None whatsoever. All manner of plant technologies are described—like photosynthesis, for example—yet never are the poignant processes described in terms of an active intelligence being uttered and expressed through DNA. Plant life is portrayed as consisting of brute facts devoid of intelligence, design, and purpose, no connection being made between the emergence of constructive DNA and the contextual configuration of Nature that elicits such nifty construction. This epistemological failure is despite the fact that each and every exhibit within Kew's Evolution House, each and every evolutionary innovation mentioned therein, is testimony to natural intelligence as it learns and excels in the art of sense-making. Indeed, the whole of Kew, as with any green surface, is brimming with speciated botanical intelligence. So rich is this organic intelligence that individual botanists at Kew, like their zoologist counterparts, can spend their whole careers embroiled in the study of a single genus or even single species.

It seems high time then that the human ego, coveting the phenomenon of intelligence solely for itself, take a step back and accord unto Nature what Nature deserves. Whether you call it Nature, the Cosmos, or the Universe, let us admit that it is an intelligence-driven, intelligence-wielding, and intelligence-generating system. The bottom line is that we and all other organisms, plant or otherwise, are spun of this intelligence, from the double helices of our DNA to the confluential cellular orchestrations brought to bear by these helices. If we cannot grant that the tree of life is intelligently composed and intelligently self-conducting, if we cannot admit that life is a natural technology crafted by a naturally intelligent Universe, then sophisticated intelligence will be limited to us alone, a zealous anthropocentric view if ever there was one. Something surely has to change in our paradigmatic take on life. But if the reader still harbors some degree of uncertainty as to the reality of natural intelligence, then our next chapter should help to allay this. It is time to detail the awesome nanotechnological intelligence that underlies our each and every breath.

CHAPTER 8

e

SYMBIOSIS

Making Sense Together

Consider, if you would, the cow. Seemingly no more than a large lumbering beast playing a major role in human culture in terms of the meat, milk, and leather it affords. Since there appears to be no evidence that our prehistoric ancestors ever came across wild stampeding herds of cows (in the manner of belligerent buffalos or moody migrating wildebeest), reason suggests that the modern mild-mannered cow is the result of centuries of artificial selection. At any rate, the fact of millions of cows currently grazing the world's green surfaces pays homage to their utility. From burgers to T-bone steaks to strong leather to the milk in our tea and coffee, it is the global farming of cows that has enabled this rewarding state of utilitarian affairs to take hold.

Although there are urgent ecological reasons to reduce widespread beef farming (because of methane production, habitat destruction, etc.), I bring the reader's attention to bear upon cows at this juncture because this chapter deals with the naturally intelligent process of symbiosis. As we shall eventually see, the cow offers us a glimpse of a particularly striking example of symbiosis without which cows would simply not exist.

183

Symbiosis is assuredly one of the great hallmarks of natural intelligence, a glaring neon sign on the new paradigmatic highway as it were, and certainly worth examining in detail, since it is such a fabulous example of the sort of emergent biological orchestration that natural intelligence is so proficient at implementing. In particular, *mutualistic* symbiosis—wherein two organisms work together for reciprocal benefits—can be so brilliant, so sophisticated in nature, that only the concept of natural intelligence begins to serve it any kind of explanatory justice. And since the humble cow plays host to an impressive example of symbiosis, I hope to galvanize in the reader's mind a fair degree of astonishment over the extent and sheer magic of natural intelligence in this particular bovine field of manifestation.

Before detailing the symbiosis involved with cows, a more classic example of mutualistic symbiosis can be brought to mind by considering the beautiful relationship between bees and the flowering plants that they visit in the spring and summer months. Each species—plant and insect—depends upon the other for its well-being. In an agreement co-drafted in naturally intelligent DNA rather than pen and paper, the bees come to receive nutritious food in the form of sugary nectar, while the plants achieve pollination and hence reproduction through the concurrent transfer of pollen by the bees while they collect nectar. To give the reader an idea of the extent and scale of this particular form of cross-kingdom symbiosis, note that fully 30 percent of the food we eat is directly or indirectly dependent upon the pollination effected by busy symbiotic bees. Synergy is another way to describe this kind of relationship. Synergy means that two otherwise independent entities, or systems, work together and produce useful results neither could achieve on its own.

Another familiar example is evinced between "cleaner" birds and rhinos. The cleaner birds get a free insectile meal from the rhinos they attend, while the rhinos get rid of the unwanted (parasitic) insects hitching a culinary ride on their hide. Similarly there exist small cleaner fish, which accompany larger fish in order to feed upon their parasites.

They do this professionally. It would be very easy for the larger fish to simply eat the diminutive cleaners, but they refrain, since there exists a steady state of symbiosis, an unspoken agreement in which both kinds of fish employ one another. Clearly not all fishy business is fishy. And if the reader recalls the fungus-growing ants and termites mentioned in the previous chapter, this kind of insect-fungus association is also symbiotic in nature. With mutualistic symbiosis, both parties benefit greatly from their specific interaction.

To tie in the synergistic phenomenon of symbiosis with the paradigm of natural intelligence, recall that we have been concerned with reinterpreting evolution, with seeing such a process not as blind and dumb but as a manifestation of intelligence. We have seen that Nature is configured in such a way that certain highly constructive phenomena are encouraged to manifest: phenomena like the emergence of self-organizing and self-replicating DNA, as well as the sensibly changing structure of DNA in light of the sensible context in which it finds itself. It is precisely this sensible context that allows genetic DNA to arise and then clothe itself with more and more refined biological systems that come to reflect or translate Nature's inherent sense and sensibleness. Driven by this sensible contextual property of Nature, DNA, bio-logic, and behavior become more and more complex as more and more of Nature's eminent sensibleness and lawful intelligibility become "known" and utilized. Mutualistic symbiosis can therefore be viewed as a sublime extension of this principle whereby two organisms, by uniting parts of their biology and their behavior, can make *more sense together*— a proficient cooperative team effort, as it were.

So what's the deal with the cow? Is the cow's role in human culture indicative of mutualistic symbiosis? Well, while it is true to say that we humans benefit in a big way from cows, in the long run cows generally get slaughtered, and therefore mutualistic symbiosis cannot really be involved in the equation. While the cow might, in the care of a loving organic farmer, have a pleasant carefree life, its imminent death and high-tech processing into organic sausages or whatever ensure that this is in no way a classic mutualistic relationship. It is rather the case that the arrangement is essentially parasitic, in that we humans parasitize cows (just as certain insects parasitize rhinos). We live off cows purely for our own selfish gain. Any benefit to any farmed cow during its short lifetime is beside the point and purely temporary. At the end of the deal, the cow gets "done up like a right kipper" as they might say in the UK (well, "a right steak" to be more precise). Such is the life of the cow at the hands of man (apart from in India, of course, where cows are considered sacred and are allowed to roam freely). This is in stark contrast to our pet cats and pet dogs, who *are* involved in a kind of mutualistic symbiosis with we humans. In return for food and shelter, pets give us companionship, the excuse to go for long walks, and other lasting, if not unusual, rewards. And we don't generally eat our pets at the end of their stay. Then again, we often castrate them, so pet-based symbiosis is a moot point.

THE INNER COMPLEXITY OF COWS

So where on earth is the symbiosis involved in the life of a cow? The answer is *inside* it. To be precise, the symbiosis in question takes place within the cow's gut and, as we shall see, in each of its cells to boot. Let's start with the symbiosis taking place in the gut of every cow alive at this moment. As we all learn from a young age via picture books, cows spend most of their time chomping away on grass. What we don't learn in such books is that the cells of which grass is constructed, as is the case with other leafy vegetation, contain cellulose. Cellulose is a mighty tough substance. In itself this is a good example of botanically based intelligence, cellulose being a fine way to protect and support the leaves of vegetal organisms (we should also note the intelligence demonstrable in the capacity of grass to withstand "upper" damage, its essential parts being belowground). In any event, no matter to what extent grass is chewed, most of the hardy cellulose within it will remain intact. Which means that it's no good having a bellyful of fresh and potentially nutritious grass if you cannot digest the stuff. Enter symbiotic bacteria. The millions of intestinal bacteria living in the warm interior of a cow's belly are just what the cow needs, for they are able to break down cellulose (i.e., they can handle and *make sense* of cellulose). Once broken down, the nutritious residue can be absorbed and be made sense of, or be handled, by the rest of the cow's digestive system. The cow subsequently gets fat and we get fat off the cow (incidentally, the symbiotic fungi cultivated by termites likewise serve to break down wood cellulose for the benefit of the termites).

The particular cellulose-dining intestinal bacteria so crucial to the cow's survival (and the livelihood of farmers) are actually found in *all* ruminating herbivores. Indeed, most ruminating herbivores have even evolved special stomach compartments to house the bacteria. This is highly astonishing when you think about it. What this widespread arrangement represents is an all but invisible system of symbiotic wizardry, so essential that if all these gut-dwelling microbes were to vanish overnight, the entire food chain would collapse. Every species of ruminating herbivore—and this includes sheep, goats, antelope, deer, and so forth—would starve to death, since without their bacterial partners they would be unable to digest plant food. If herbivores were to die out, carnivores would follow suit. Needless to say, the biosphere as we know it would be in utter turmoil.

These hidden hordes of intestinal bacteria—known as methanogens since they produce methane (a significant greenhouse gas)—abide in gargantuan amounts. Although we might never see them, we can be assured that they do exist and that they do aid the survival of herbivores. That these bacteria are symbiotic and derive benefit from their herbivore hosts is indicated by the fact that they are anaerobic, which means they can survive only in an oxygen-free environment like that of a mammalian digestive system (such bacteria originally evolved long before oxygen became a major component of the atmosphere). In return for their breaking down of cellulose, the bacteria are thus rewarded with a homely place in which to live and thrive. If exposed to oxygen—and the air we breathe contains some 21 percent oxygen—these bacteria die. And if they die so does the cow. The mutualistic partnership is therefore a brilliantly executed symbiotic marriage of convenience. Either organism, herbivore or its intestinal microbial host, makes sense only in the light of the other. Both are entwined with one another as closely as are individual strands of rope fiber.

From this single example, it is clear that symbiosis has far-reaching consequences. Without it, evolution would have fewer directions in which to maneuver, fewer degrees of creative freedom to explore and build upon. If we conceive of any species of creature as being the embodiment of one particular realized system of intelligence, we can see that in certain situations two systems of intelligence can come together and then, between them, make even more sense. Any one organism already embodies one set of sense-making strategies, yet two organisms acting in concert can *maximize* their sense-making potential. This is exactly what gut-dwelling bacteria and cows achieve together. By joining and orchestrating their bio-logic to such an intimate degree, they ensure each other's survival. Indeed, split apart these two symbiotic partners and neither is able to make as much sense in most other contextual situations. A cow would not be able to make sense of the grass it eats (the cow will be unable to digest the grass), and the bacteria, should they find themselves inside, say, a fungus instead of a cow, or inside the cow's liver instead of its gut, would likewise have little chance of making any sense. But put cow and bacteria together in the right way and both species prosper.

Another pertinent example of bacterial symbiosis concerns piglets. Baby pigs need a particular enzyme with which to digest their mother's milk. We have all seen how piglets zealously attach themselves to their mother's teats in order to obtain nourishment and grow. Yet they can digest the milk only so long as they have inside them a particular enzyme. And this enzyme is provided by symbiotic bacteria. Once again, smart symbiosis provides and allows life to flourish. Extensive colonies of symbiotic bacteria, each with a specific function, become woven up with mammalian biology to such an extent that neither species—microscopic bacteria or macroscopic mammal—can survive without the other. Should one chance to purchase a bio-yogurt containing certain bacteria deemed beneficial to digestion, one might like to reflect upon the fact that such yogurt contains countless unseen unicellular organisms chockablock with digital DNA code whose symbiotic potential ensures that they fit nicely into the living environmental context that is the human digestive system. As Queen sang: "It's a kind of magic."

Bearing in mind the role of mutualistic insects (like the bee) in ensuring the reproduction of flowering plants, along with the symbiotic microbes just noted, it is striking to consider not only how crucial is the role of mutualism in the web of life but also how far such a cooperative process can go. Species can become so intermingled in their life patterns that it becomes hard to tease them apart. This highlights the fact that individual genes—considered as simply selfish entities by neo-Darwinian dogma—must take into account the genes of other species if they are to make sense. In other words, not only must genes code for useful structures able to make sense of the natural environment, they must also be able to make sense within the context of *all the other genes they are likely to interact with.* In the case of the cow, there are clearly genes involved with its instinct to eat grass, for instance. These genes become fully sensible only in the presence of the grass-digesting symbiotic bacteria inside the cow. And the genes in these bacteria that code for grass digestion likewise make sense only if the bacteria find themselves inside a grass-providing cow.

Biology is thus to be understood as an interweaving system, a complex integrated network of relationships that depend upon one another for their survival. Organisms, their genes, and their gamut of behaviors are all cross-referenced into a self-sustaining network. As many biologists are fast becoming aware, Nature is not simply red in tooth and claw, nor can organisms be understood in terms of individual genes isolated from the genes of other species. On closer inspection, and by becoming sensitive to the full and multilayered grain of the biological picture, the biosphere reveals itself as being built up from cooperative endeavors, mutualistic symbiosis allowing great and enduring patterns of sense to be made.

Consider the fact that *over three-quarters of all plants have underground symbiotic fungal partners* and one soon appreciates the sheer scale and importance of mutualistic networking. The plants obtain rare minerals from the fungi in exchange for photosynthetic food. Forests are thus forests of unseen symbiosis, the fungal symbionts having likely facilitated the groundbreaking transition of plant life from the sea to dry land. Cooperation is clearly just as important in the evolution of the tree of life as is out and out competition. Alas, this was not fully appreciated in Darwin's day, the popular phrase at that time being "the survival of the fittest." It is now clear that fitness and the ability to reproduce depend upon sense-making, and that sense-making often involves components and systems working together, whether these parts happen to be genes or organisms. Nature decrees not simply the survival of the fittest, but the survival of that which makes sense.

BECOMING AS ONE

Symbiotic gut-dwelling microbes were detailed merely as an aperitif to whet the reader's conceptual appetite. If you suddenly realized the importance of the symbiotic bacterial activity silently (until wind is broken . . .) transpiring in mammalian digestive systems, our next example is probably the greatest of all feats of symbiosis. Or at least it represents the greatest example that we are wise to at the current time. (Since much of the magic of symbiosis has become fully recognized by science only in recent years, there are certainly more instances in the same league that have yet to be discovered.)

It was indicated earlier that the cow was once more involved in this next form of symbiosis. Indeed it is. And like the previously discussed bacterial symbiosis, this other form of mutualistic symbiosis takes place not just in cows but in us humans too (all animals, actually). The symbiosis also involves bacteria, only this time they are so intimately involved with animals, so finely interwoven, that they have become for all intents and purposes quite literally *innate* parts of animals.

One can already divine the trend. First, two organisms, or two living manifestations of organic intelligence, come together and aid one another in survival and sense-making. Such mutualism, such a new and emergent concrescence of intelligence, can then become so deeply substantiated that each organism depends upon the other for the *whole* of its life. The symbiotic contract has no expiry date. Taking this operating principle further, we find adroit organisms like lichens. Abounding on the surfaces of rocks, lichens look like a single organism, like licks of colorful living paint able to withstand extreme changes in temperature and humidity for hundreds of years (some can actually live for four thousand years!). On closer inspection it emerges that lichens consist of two organisms—a fungus and an alga—locked tightly together in a living embrace, each depending upon the other for its survival, so much so that science has yet to come to terms with how lichens manage to replicate themselves in toto. In the case of lichens, the symbiotic contract involves almost total integration as well as an indefinite duration.

Here we reach the border between differentiated organisms and a new and incipient species. For it could be argued that the lichens of the world represent actual holistic species; that the fungus and the alga are so bound up with one another that a single higher-level organism has emerged. This surely depends upon the eye of the beholder. After all, we call them lichens and have documented more than eighteen thousand species. When we see them on rocks we don't immediately sense that two organisms are present. Lichens are on the verge of complete synergistic integration, almost as if they were offering us evidence of some transitional phenomenon in the field of biological organization in which natural intelligence excels.

In a similar fashion, the green hydra, a freshwater creature consisting of a long tubelike body with tentacles at one end with which to sting and catch prey, acquires its name from the fact that it has evolved a symbiotic relationship with a green alga. This alga lives inside the hydra, where it receives protection and motility, the latter allowing it to get light from the water surface beneath which the host hydra swims. In return for harboring the alga, the hydra receives oxygen and any surplus carbohydrates from its internal photosynthesizing partner. When the hydra reproduces via budding, the alga is concurrently reproduced. As with lichens, this kind of symbiotic relationship represents fine-tuning whereby different strands of already realized smart bio-logic weave themselves together to produce a new whole, which, in this case, is so complete that between them the hydra and the alga have essentially formed a new organism.

There are many examples of this kind of inventive tight-knitting. Corals for instance. Corals are not underwater plants but essentially colonies of tiny invertebrate animals—known as polyps—whose aggregated skeletons shed over innumerable years form reefs. Corals contain algae within their translucent saclike bodies, which explains why they are found in shallow waters near the light-rich surface of the ocean. The algae, of course, photosynthesize, 80 percent of their "takings" being extracted as "rent" by the polyps that harbor them. In return for their rent, the algae get a protective niche in which to flourish as well as free meals of nitrates and phosphates excreted by the polyps. Huge coral reefs are therefore colorful underwater mosaics fashioned throughout with no-nonsense, no-waste symbiotic synergy.

There is even a common sea slug on the coast of Australia that eats coral polyps and somehow manages to separate out the algae, which it employs for its own metabolic ends. Once separated, the algae are put to service in specially evolved long tentacles that branch out from the sea slug's gut. The sea slug then stimulates the algae so that they proliferate in vast numbers. After this green-fingered success, the sea slug *does not need to feed again*. Receiving adequate sustenance from its internal algae, the sea slug can relax and occupy itself with whatever it is that sea slugs do in their leisure time. Although it would appear that the sea slug is getting the lion's share of the symbiotic benefit, it is equally possible to conceive of the algae as being just as favored in the interaction in terms of reproductive advantages. In any case, if this form of "inner garden" symbiosis were enacted by humans—we could imagine, say, genetically altered humans who no longer need to eat because they have green photosynthetic skin—it would represent the grandiose stuff of science fiction, yet it is precisely what these sea slugs have achieved.

Two-way lichens, corals, sea slugs, and green hydra embody an essentially logical process, a logical development of natural intelligence as it constantly evolves. If two organisms can work together in order to maximize resources and aid one another's survival, then natural intelligence will ensure that their symbiotic union comes to pass, since natural intelligence is precisely in the business of manufacturing sense, this sense being wrought through bio-logic. Clearly, by working together in harmony, two organisms can make more sense than by functioning on their own, just as, say, our computer systems can function together with telecommunication systems. Although built and designed for seemingly separate functions, after the advent of the modem, computers suddenly became intimately married to phone networks, resulting in a form of synergistic telecommunicational "symbiosis" upon which the Internet was able to flourish.

BREATHTAKING INTELLIGENCE

The boundary between two separate organisms and one greater organism is really and truly crossed in the example to which this long foray has been leading. I here refer to *mitochondria,* those parts of animal cells that supply energy. Mitochondria are organelles that are found in all our cells and supply the energy needed for metabolic processes. Place any one of your trillions of cells under a powerful-enough microscope and you will be able to spot mitochondria. Query biologists and they will inform you that mitochondria are the powerhouses of cells, the energy centers, just as a power station is the energy center of a city.

In fact, mitochondria are responsible for the fact that we *breathe*. We all breathe. Yet how many of us know exactly why we breathe? For oxygen? Of course. But why do we need oxygen? Enter mitochondria. These active structures make intelligent use of oxygen by employing it in certain chemical reactions in which energy is released from absorbed food. Oxygen is good for this purpose because it is so reactive. Its use in the controlled release of energy from food can be considered a slow form of combustion achieved via the activity of mitochondria. In other words, mitochondria are indeed like busy power stations. By combining the oxygen we breathe with the food we eat, they provide energy to their host cells, which can then function and support the workings of the organism of which those cells are a part. It should be clear at this point that mitochondria, through their energy-production skills, keep *all* multicellular animals alive. The most intelligent transformation of all the food we eat and all the air we breathe comes to light via the ceaseless activity of mitochondria.

The really intriguing thing about mitochondria is that *they have their own DNA*. Although this has been known for quite some time, it was not until the late 1960s that biologist Lynn Margulis resurrected the radical thesis that mitochondria were once *free-living bacteria* that, through symbiosis, have become permanent residents within multicellular organisms. Such a fact would explain why mitochondria have their own DNA, distinct from the DNA stored within the nuclei of the cells they occupy. Not many believed Margulis at the time. Biologists were prone to snicker at Margulis, thinking of her as some young upstart scientist whose fertile imagination had gotten the better of her. It seemed too outrageous and indicative of a kind of organic engineering wizardry never dreamed of. No surprise then that more than a dozen scientific journals refused to publish her first major paper on the subject. But over the past few decades Margulis's theory—called *endo*symbiosis—has become accepted by most mainstream biologists. Even Richard Dawkins has finally acknowledged Margulis's work, referring to her as the high priestess of symbiosis—although it must be said that he still insists that symbiosis is no more than "selfish cooperation," an oxymoron in which one senses a last desperate attempt to hold on to an outdated selfish gene metaphor.

What all this means is that way, way back down the evolutionary tree of life, before it branched out into the major kingdoms of life, certain free-living bacteria (the very first organisms) joined in symbiotic union with other unicellular creatures and became so symbiotically entwined that eventually multicellular creatures (entirely new *kingdoms*) were born. Both the nucleus and the mitochondria of animal cells are believed to have arisen from endosymbiotic processes. As Margulis and Sagan put it:

> All cells have a nucleus or do not. No intermediates exist. The abruptness of their appearance in the fossil record, the total discontinuity between living forms with and without nuclei, and the puzzling complexity of internal self-reproducing organelles [e.g., mitochondria] suggest that the new cells were begotten by a process fundamentally different from simple mutation or bacterial genetic transfer.[1]

In other words, previously independent bacteria become suddenly and dramatically linked as one through a process of coevolutionary symbiosis, each species aiding the other in its survival. The hypothetical scenario involves one bacterium enveloping another. But instead of digesting it, a form of synergy manifests in which a mutualistic partnership is borne:

> With time these populations of coevolved bacteria became communities of microbes so deeply interdependent they were, for all practical purposes, single stable organisms . . . Life had moved another step, beyond the networking of free genetic transfer [bacteria swap DNA freely] to the synergy of symbiosis. Separate organisms blended together, creating new wholes that were greater than the sum of their parts.[2]

It is not just mitochondria that provide us with evidence of seamless endosymbiosis. Chloroplasts—those fabulous structures that exist in plant cells and enable photosynthesis to take place—are likewise thought to have once existed as free-living bacteria that gave up their "roaming" in order to take up permanent residence inside other bacteria. On its own merit, photosynthesis is remarkable, yet if one dwells upon the fact that the origins of the photosynthetic machinery within plants lie in once separate creatures, one's degree of amazement over the creative potential of natural intelligence shoots off into infinity.

As a simple example of how compelling this endosymbiosis hypothesis is, it has been noted by geneticists that the DNA found in chloroplasts (which is distinct from the DNA in the nuclei of plant cells) looks highly similar to the DNA found in cyanobacteria, which are bacteria able to photosynthesize (it is believed that cyanobacteria *invented* carbon-dioxide- and water-based photosynthesis; they preceded multicellular plants). Which implies that once upon a time chloroplasts, like mitochondria, really did have some sort of free-living bacterial ancestor, probably a cyanobacterium.

All this fits well into a neat theory, endosymbiosis emerging as a truly astonishing manifestation of natural intelligence, a phenomenon that human culture is actually pretty familiar with although in a different form. As Margulis and Sagan point out:

> Human religion and mythology have always been full of fantastic combinations of creatures—the mermaids, sphinxes, centaurs, devils, vampires, werewolves, and seraphs that combine animal parts to make imaginary beings. Truth being stranger than fiction, biology has refined the intuitively pleasing idea with its discovery of the overwhelming statistical probability of the reality of combined beings. We and all beings made of nucleated cells are probably composites, mergers of once different creatures.[3]

This kind of tight endosymbiosis can occur quite abruptly. Margulis has noted biological research that, after ten years, yielded an amoeba that could survive healthily only with a certain species of bacterium living inside it. Rapid evolutionary developments like this drive home the point that whatever works, whatever makes sense, becomes selected by Nature. Different strands of naturally intelligent bio-logic will, where possible, unite, thereby allowing evolution to branch out and achieve breathtaking patterns of symbiotic beauty. It's a technique really, a potential of natural intelligence guaranteed to happen given the right circumstances and enough time.

Considering that we ourselves are multicellular creatures and that all of our cells contain mitochondria, just as most plant cells contain chloroplasts, it is evident that natural intelligence has, through endosymbiosis, transcended itself and allowed the plant and animal kingdoms to take hold and flourish, a rather staggering insight recognized and properly appreciated by mainstream science only in the past few decades (even *Star Wars* creator George Lucas has picked up on endosymbiosis by basing endosymbiotic "midi-chlorians"—the source of the Force—upon mitochondria). According to the paradigm of natural intelligence, endosymbiosis allows sensible bio-logic to become more firmly established and even expand the potential for further evolution. Moreover, it is a natural and predetermined consequence of the way Nature is configured. Both endosymbiosis and other forms of symbiosis happen again and again, quite naturally, and are indicative of the persistent creative flow of natural intelligence. Life is truly a miracle of nanotechnological craftsmanship. That Queen song comes to mind yet again.

THE EVOLUTION OF COOPERATION

To further highlight the role of cooperative symbiosis within evolution, studies by political science professor Robert Axelrod using computer simulations of evolution found that digital entities forced into "meeting" one another would invariably be favored and selected if they cooperated with one another. In other words, when simulated creatures mingle with one another and can either cooperate or exploit one another (over resources and such), then cooperation (i.e., mutualism) will *always* be the most adaptive response if the interactions are engineered to be consistent and permanent.

By mathematically *proving* this principle in successive computer simulations, Axelrod was basically demonstrating the beautiful logic of natural intelligence—that evolution will tend, in the course of time, to produce systems in which the components cooperate in some way. We can see this eventuality most markedly in both endosymbiosis (i.e., in mitochondria and chloroplasts) and the working together of individual genes within a single organism. The key is once again sense-making. In the long run, it makes sense to work together if that is possible. This is why the tree of life is full of interconnected systems of sense-making. In the case of multicellular organisms like plants and animals, symbiosis of one form or another makes it all possible. Likewise with the individual genes within any genome, which must perforce cooperate in a sensible and intelligible manner with their fellow genes.

FRUIT

Full of Symbiotic Juice?

Symbiosis has also been invoked as an explanation for the evolution of *fruit*. It seems that the genesis of seed-bearing fruit as a botanical mechanism for reproduction was a technique that arose quite abruptly some forty million years after the emergence of flowering plants. The radical idea is that the genes of certain species of fungus became mutualistically incorporated into plant genomes with the effect that proto-fruits were borne, enabling both plant and fungus to reproduce more successfully. The idea calls to mind *galls,* protrusions observable on many plants. As well as being produced by insects and bacteria, some galls are produced by fungi whose DNA is reacting with the host plant's DNA to produce distinctive and often bulbous gall structures. And some galls *do* look a bit like fruit. Indeed, some galls look remarkably healthy, as opposed to looking like some disease symptom. Galls on the backs of leaves can actually appear "ripe" and to be some sort of complex extension of the leaf, suggestive of function. Similarly, common oak galls—caused by parasitic wasps—can look like small fruit. They even turn from green to brown at the end of the summer. Thus, the gall tissue, which may result from an interaction between fungal DNA and plant DNA, might not necessarily be a destructive thing; rather such tissue can become the starting point for a new symbiotic path for natural intelligence to explore.

Given the oft-constructive appearance of galls, the notion that the first-ever fruits were borne of a plant/fungus symbiosis is compelling. It seems that there are clues all around us to the techniques employed by natural intelligence. A biologist specializing in symbiosis is rather like an investigator on the trail of a pantheistic craftsman, often locating all manner of signs and indications of its existence, galls being, perhaps, one tantalizing clue of how natural symbiotic wisdom can emerge.

ENDOSYMBIOSIS AND THE BRAIN

Margulis and Sagan have even gone on to speculate about the possible endosymbiotic nature of microtubules, which are tiny protein structures found within nerve cells (i.e., neurons) as well as inside certain bacteria. If the microtubules in neurons were indeed found to be of microbial origin (derived from spirochete bacteria), then the human mind— presumably bound up with the organized firing patterns of billions of neurons—would itself be dependent upon endosymbiotic bacteria. Margulis and Sagan ask:

All our favorite inventions were anticipated by our planetmates; why not our thought? If bacterial "cold light" (bioluminescence) preceded electric lights by 2,000 million years, if the protist *Sticholonche* propelled itself with microtubular oars through the Mediterranean long before Roman galleys rowed the same waters, if horses roamed the plains, whales the seas, and birds the skies long before the traffic of cars, submarines, and airplanes, is it so farfetched after all that bacterial symbionts created biospheric information pathways as important as quantum mechanics or the theory of relativity? In a sense we are "above" bacteria, because, though composed of them, our power of thought seems to represent more than the sum of its microbial parts. Yet in a sense we are also "below" them. As tiny parts of a huge biosphere whose essence is basically bacterial, we—with other life forms—must add up to a sort of symbiotic brain which it is beyond our capacity to comprehend or truly represent.[4]

What Margulis and Sagan go on to suggest is that the microtubules within the hundred billion or so neurons in the human cortex are the remnants of once free-living bacteria, in the same way that mitochondria and chloroplasts are the remnants of bacteria. Since thought, perception, cognition, and memory are all believed to be mitigated by the firing patterns of neurons within the brain, the implication is that endosymbiotic bacteria have, ultimately, aided in the emergence of cortically embodied sentience. If this proves to be true, then it will be yet more startling evidence of the almost ironic superiority of life's intelligence—expressed through symbiosis—over human intelligence.

Considering mitochondria, chloroplasts, and, if Margulis and Sagan are correct, microtubules, symbiosis is responsible for the emergence of not only multicellular animals and plants but also our thoughts to know this. In each case, endosymbiotic bacteria have facilitated the formation of complex macroscopic architectures. The symbiotic partners involved are akin to strategic segments of a living jigsaw that were always poised to come together. And who, in looking, say, at their hands, can truly grasp the fact that underpinning the massive conglomeration of cells tightly bound into an enduring macroscopic architecture there reside once free-living bacteria, or at least the mitochondrial remnants thereof? And that this experience of astonishment is due to a neuronal brain whose structure incorporates similar symbiotic ancestry? This is symbiosis-driven ingenuity with a vengeance.

GLOBAL FORMS OF SYMBIOSIS

Although mutualistic symbiosis is generally thought of as being a phenomenon in which two organisms are physically entwined in some way, it might well be that more expansive forms of symbiosis exist. Which is to say a form of symbiosis between species living far away from one another. The idea is that although separated and living in completely different environments, two or more different species might still have a relationship in which their behavior, or the *indirect consequences thereof,* aids one another's survival. The gist of the idea is that the environmental consequences of each species' behavior exact mutually beneficial effects. It would be a bit like the population of London acting in some way that enhanced the life (and thus reproduction) of people in New York. In consequence, the population of New York might enhance the life of Londoners. So although both groups are not physically touching one another, nevertheless their actions are mutually beneficial and become positively selected over time.

This admittedly extravagant-sounding notion can be explored and weighed up if we consider the following information. Over the last few decades James Lovelock's Gaia theory has received much attention. According to Lovelock (and Lynn Margulis, who helped develop Gaia theory), the entire biosphere is best understood as a single cybernetic system in which all organisms feed-back upon one another and influence one another and in which the oceans and the atmosphere are similarly important parts of the system. This must be true to a certain extent, since we have established that no one organism is truly separate from all other organisms but rather all and sundry are part of an interconnected contextual web. In this light, Gaia theory suggests that key biospherical parameters such as the gaseous constitution of the atmosphere, ocean salinity, and ambient global temperature are all actively maintained by life through homeostatic feedback mechanisms. It is thought by proponents of Gaia theory that such forms of global homeostasis in which optimal life conditions are maintained are an *emergent* property of planetary life. Mathematical models have been developed that show how cybernetic control mechanisms *can* and *do* emerge within certain systems. Just as a homeostatic temperature control mechanism can emerge from the collective interaction of social insects such that a hive maintains a healthy temperature, so too might global forms of homeostasis emerge from the combined interaction of various species of organisms.

To put this in the simplest way possible, the theory implies that, in the long run (and that can mean millions of years), only those species and lineages of organisms whose effects upon the biosphere are either neutral or beneficial in some way will prosper. By beneficial, I mean that the effects serve to keep some environmental parameter like temperature at an optimum level. The biosphere then emerges as a predominantly cooperative system in which conditions for *bio-logic in general* are optimized. Although parasites and "cheaters"—organisms with life strategies that might positively harm other organisms or even upset global parameters—can still flourish, the most important principle will nonetheless be one of cooperation. Genes, genomes, bodies, and species that *can* cooperate and *can* ensure one another's survival in some way will prosper, regardless of the distance between them. Hence we live in a biosphere whose global properties are, or at least would appear to be, kept at an optimum level for life. Indeed, when one considers the sheer turnover, or flux as it is more properly called, of atmospheric gases like oxygen and carbon dioxide, it is clear that some sort of vastly distributed control system is in operation, a control system made up of countless interacting organisms. Just as the human organism constantly imbibes and excretes gases like oxygen and carbon dioxide, so too does the entire biosphere constantly produce, circulate, and deposit staggering amounts of these two metabolic gases. The end result is the maintenance of the carefully balanced atmosphere so crucial for life.

In considering the weblike nature of life in which all parts influence all other parts, global symbiosis will come into play when the environmental impact of one species serves to promote the survival of another species (as it did with our imaginary New Yorkers and Londoners). In turn, the latter species will aid the survival of the former. Although this interaction may take place over a relatively long period of time and over vast distances, all that matters is that a self-sustaining mutualistic loop emerges. This, of course, is exactly what mutualistic symbiosis is all about, but in this case we are speaking not of two organisms tightly coupled together as we witness with, say, lichens, but of two or more species coupled together in some global manner. In other words, a phenotypically extended form of symbiosis.

The best way to conceive of this sort of large-scale symbiosis is to consider the following phenomenon, which has been much discussed by Lovelock. It is intriguing to say the least, and worthy of committing to memory. The phenomenon concerns the environmental impact of coccoliths, which are a type of algae that exist in gargantuan numbers in the ocean, forming what are called algal blooms, which can be hundreds of miles in length and breadth. From the vantage point of an orbiting satellite, the oceanic blooms look like immense spills of green paint. Unlike dark spills of oil, the green algal blooms are benign, consisting of trillions of photosynthesizing coccoliths, each one of which is no more than two-millionths of a centimeter in length. As individuals they are invisible; en masse they show themselves.

It has been firmly established that apart from shedding calcium-containing shells, which lead to deposits of limestone on the ocean floor (that's what the famous white chalk cliffs of Dover are made from), coccolith blooms also produce and release large quantities of a gas called dimethyl sulfide as part of their collective metabolism. This gas rises into the air and serves to promote the formation of clouds. This happens because raindrops—of which clouds are made—need a surface upon which to form. Just as water vapor can condense on the surface of a window, so too will it condense up in the atmosphere if there is a surface available. Dimethyl sulfide provides a surface for water to condense upon because this gas reacts with the oxygen in the atmosphere and leads to micro-droplets of sulfuric acid. These micro-droplets, or nuclei, then act as surfaces upon which water can form into raindrops.

The large-scale consequence of dimethyl sulfide production is therefore cloud formation. By releasing huge amounts of dimethyl sulfide, blooms of coccoliths actually cause white fluffy clouds to arise. In itself, this is remarkable. But what effect will this cloud formation have? The possible consequences are many. The first notable consequence to consider involves ambient temperature. Since clouds reflect sunlight—called the *albedo effect*—then a large cloud cover during the day can reduce temperatures. And lower temperatures are likely to be associated with less reproduction in the coccoliths. Fewer coccoliths will lessen the production of dimethyl sulfide. Less dimethyl sulfide and therefore fewer clouds. Fewer clouds means a diminished albedo effect and higher daytime temperatures. Higher temperatures mean more coccoliths. And so on. Here we can see a possible form of global temperature control in action. But there may well be more to it than that. Another consequence of dimethyl sulfide release is that it facilitates the sulfur cycle. The sulfur within the gas will, through its journey within clouds and raindrops, be deposited onto the land, and this will be of great benefit to the plants and animals living on the land, sulfur being an absolutely vital commodity needed by all organisms. Indeed, the sulfur transmitted from the oceans to the land via the algal production of dimethyl sulfide gas is believed to represent an important part of the sulfur cycle.

Enhanced by this arrival of sulfur, plants and most other land organisms grow more efficiently and break down rocks at a greater rate. Rock weathering is also important for living organisms because it releases essential minerals. It has been shown that in the presence of organisms rocks will be broken down one thousand times faster than in sterile non-life conditions. Large quantities of minerals released in this way will, apart from being incorporated into local plants, be washed into rivers and, in consequence, transported to the sea. So, the increased vigor of land life owing to the arrival of sulfur originally produced by oceanic algae eventually leads, willy-nilly, to more minerals being conveyed to the ocean. And more minerals in the ocean is *exactly what algae such as coccoliths need.* Thus the algal coccoliths prosper due to the long-term effects of the dimethyl sulfide that they produce, while the plants on the land prosper because of the long-term effects of the weathering of rocks that they cause. An exquisite life-affirming circuit is closed.

Viewed on a biospherical scale we can thus see that a real kind of symbiosis exists between oceanic algae and their botanical cousins on the land. The long-term consequences of each species behavior are mutually beneficial. It is this sort of reciprocal feedback scenario (similar in kind to the *autocatalytic sets* we will meet in the following chapter) that can be viewed as being a form of global symbiosis. In the long run such a system makes great sense, and this is precisely why Nature will, over long periods of time, selectively establish such an elegant scheme. Whereas an organism will be selected by Nature if it behaves in such a way as to enhance its own survival, any mutually beneficial activities that emerge between groups of organisms—whether widely separated or closely knit—will also be selected. In the case of widely separated species, the evolution of long-distance forms of symbiosis may take a great deal of time to become realized, but this is exactly the sort of eminently sensible orchestration that natural intelligence will implement through its constant evolutionary growth. Once again, it is a case of the survival of that which makes sense.

It is entirely possible that many other similar systems to the one discussed have also evolved. Indeed, given that these forms of global symbiosis make good sense for all the organisms involved, their emergence is ensured. And the real miracle of course is that we conscious humans can divine all these facets of natural intelligence. By doing so, by consciously perceiving natural intelligence, we would appear to be fulfilling a new potential of Nature to know itself and more fully awaken to itself. We explore such radical ideas in the following chapters.

CHAPTER 9

ATTRACTORS AND THE EVOLUTIONARY EMERGENCE OF MIND

Until now, our conception of natural intelligence has been built firmly upon extant scientific data. From the facts at hand, we have seen how the evolution of the tree of life represents a natural learning process. This must be so since all organisms represent elaborate biological systems *wise* to their environment and able to actively make sense of their environment. This is what evolution does—evolution delivers not simply an accumulation of DNA modifications but rather a series of eminently sensible life-enhancing modifications. We also saw that it is the intelligently configured context of Nature that affords the learning by providing the sense and order (or specific information) that the evolving tree of life feeds upon. Thus, we can conclude that Nature is everywhere imbued with intelligence, from its lawful configuration to the biological systems (and even ecological systems) that evolve in response to this lawful configuration.

We have also established that while the manifest facts of life are indisputable—like the existence of the genetic code and the speciating systems of bio-logic engendered by that code—interpreting these facts in terms of natural intelligence is, for one reason or another, discouraged. Indeed, describing life as a technology or an intelligence is in stark contrast to the merelyistic and reductionistic discussions of life currently all the rage among evolutionary biologists. Nevertheless, although difficult to test as such, the hypothesis, or paradigm, of natural intelligence certainly serves to explain the tree of life's existence far more aptly than does any reductionistic creed. And since one definition of the word "hypothesis" describes it as an explanatory framework for a group of observed facts, then it is safe to say that natural intelligence represents a sound hypothesis resting upon a firm ground of evidence, the same firm ground, in fact, upon which reductionistic hypotheses are based. Only the attitude and perspective differ.

At the end of the day, it probably all boils down to language. As I have stated elsewhere, we seem to reserve words like intelligence and technology for ourselves and are loathe to expand our definitions to include life itself. We also seem incapable of separating intelligence from consciousness. In that sense, maybe the first step to accepting natural intelligence is to conceive of unconscious intelligence—that Nature is, as far as we know, an unconscious intelligence that acquires conscious aspects once nervous systems have evolved to a certain level of complexity. In any case, given its paradigmatic importance, my hope is that the study of natural intelligence will become as popular in the collective imagination as, say, the contrary study of artificial intelligence. While the popularity and media coverage of artificial intelligence are guaranteed to grow in the coming years due to the explosive growth of computer technology and robotics, an emerging awareness of natural intelligence is arguably more worthy of cultivation. This is especially so when you consider the extent to which scientific paradigms shape and influence human culture. Indeed, it is not unusual for scientific ideas to eventually wriggle their way into the political arena. Providing the human race with answers pertaining to the meaning of existence, scientific paradigms can shape cultural values and social policies. This is the power of a paradigm once it has been conveyed through the educational establishment and has settled in the collective psyche. The danger, of course, is that a paradigm might be lacking in some way or is misguided (incidentally, in a personal letter, Gaia scientist James Lovelock warned me that new theories take forty years to gain acceptance!).

Our interpretation of biological evolution is assuredly of great importance, since evolution purports to explain exactly why and how the tree of life came to be. It might well be that without acknowledging the intelligence inherent within Nature, we are missing the whole point of life, rather like missing the essential plot of a book or a film. This is why the concept of natural intelligence deserves close scrutiny. If it really is a better paradigm of life, then it can certainly serve to influence cultural values and cultural behavior. The ecological healing of the planet, for example, necessitates a radical change in our relationship with the rest of Nature, a change that the paradigm of natural intelligence affords.

I can also stick my neck out here and assert that the natural intelligence paradigm can aid in the ongoing study of the genetic mechanisms at the heart of life. In particular, the notion of the *evolution of evolvability* will likely prove fruitful if we acknowledge that a cell is a manifest system of intelligence that might be able to influence its own potential to evolve. By this I mean that life may have learned how to regulate genetic changes in some as yet unknown way. Take so-called junk DNA, which makes up the bulk of genomes. This was christened "junk" because no function was found. And yet more and more evidence is emerging that junk DNA is not junk at all but has some important regulatory function vis-à-vis genetic expression. If we also speculate that there is an underlying *grammar* to protein building (which is what genes do), then it is feasible that life has learned this underlying grammar and can, by some mechanism, influence grammatically correct changes to DNA, or at least modifications that are more likely to be grammatically correct. It may sound far-fetched, but it is no more far-fetched than the DNA-error-correcting machinery already known to be at work in cells. Many things are possible with natural intelligence. I am convinced that the more carefully life is studied, the more astonishingly sophisticated life will be seen to be.

Apart from priming our minds to discover hitherto unknown subtle genetic control mechanisms, the purely metaphysical implications of the natural intelligence paradigm are likewise of great import, especially when it comes to the role of consciousness within the evolving web of life. This, in fact, is where we now turn our attention. Although much of what follows will be highly speculative, and the ground far less substantial than the ground upon which we have traveled thus far, I have endeavored to bring to the "metaphysical table" as much appropriate "solid" evidence as can be mustered. For reasons of clarity, in the following discussions of consciousness what is being referred to is the kind of consciousness that we humans happen to be blessed with, the kind that can learn about how and why we came to be. As to whether other organisms—such as dolphins and elephants, for instance—enjoy similar states of consciousness, this is an open question far too complex to be detailed here.

To kick off this continuing exploration into the diverse realms of natural intelligence, I should like to utilize another vivid thought experiment. A word of warning: At first, the reader might think I have gone barmy, for the scenario I am about to convey is truly bizarre and seems to have nothing whatsoever to do with either evolution or consciousness. Indeed, it might come across like the feverish and surreal nocturnal ramblings heard within the walls of some asylum for deranged mineralogists. Nonetheless, bear with the scenario, for it serves to warm us up to the essential subject matter of this chapter.

CAVE WORLD

Consider the formation of a stalactite hanging from the roof of an immense wet cave. Let's say that the only thing that exists in the world is this water-ridden cave and its single growing stalactite. If you like, you can imagine the water in the cave as forever circulating through evaporation and condensation. Let's call this wet imaginary place "Cave World."

Like any stalactite, our envisaged one inside Cave World is a mineral structure formed from the incessant deposition of limestone from the rivulets of water constantly flowing over it (particles of calcium carbonate are carried in the water). Over millions of years, the stalactite builds up, its growth and structure determined by the physical and chemical laws operating within Cave World. In other words, the stalactite forms itself quite naturally, a millimeter or so of deposition being added to its dimensions each year. No surprises there. All the physical and chemical events transpiring within Cave World are totally in accord with natural law. However, imagine that the stalactite now begins to grow in a strangely convoluted manner, very much different from the long, thin structure we are accustomed to associate with stalactites. Over untold millennia the stalactite within Cave World starts forming a complex shape, folding in upon itself, becoming more and more intricate with each passing century. Actually, if we look closely, a patterned structure seems to be emerging. Not just any old pattern, mind you, but some sort of architecture reminiscent of *wiring and circuitry*. Indeed, let us imagine that delicately thin tubules begin to form with an internal structure such that signals and information can be transmitted within them. Maybe electrically charged calcium ions within the limestone serve to carry information. The details do not concern us. What matters is that pale, moist bundles of tubular information-processing structures are, eerily enough, being self-organized within the growing stalactite.

The surreal nature of Cave World has only just begun. As untold aeons pass, the structure of the developing stalactite becomes so convoluted as to suggest some kind of information-processing machine, perhaps a computer. Hang on . . . It's not a pale limestone computer that is slowly forming but some kind of visual equipment! It looks like a video camera, perhaps even a HD digital one. Imagine that! Over hundreds of millions of years of sluggish particulate deposition, the stalactite has somehow managed to form itself into a fully functioning video camera complete with internal electronic wiring. Moreover, if we eagerly fast-forward through millions more years of Cave World, the (calcite) lens of this stalactite camera is not positioned downward but appears to be set at an angle. What kind of outlandish natural limestone formation is this emerging? Ah, after a few more millions of years the case is clear. A mechanical video camcorder device has indeed formed; only its lens apparatus is set at almost 180 degrees, such that it is actually focusing back on whence it came. And it even has a light source with which to illuminate its gaze. In other words, a self-reflective, self-observational system has organized itself into existence. We can imagine that the camera, positioned so, *is thereby able to examine its own form, its own nature, and is able to observe from whence it came and thus be privy to the lawful nature of the world in which it has arisen.*

THE REAL WORLD

The peculiar plot transpiring within Cave World, while insanely unlikely, is nevertheless directly analogous with, although *less* astonishing, than our own situation as conscious agents within the Universe. For it is apparent that Nature has conspired, via the evolutionary process, to create nervous systems and brains endowed with the capacity to consciously apprehend their roots. Each one of us, each of our naturally formed consciousness-conducive cortices, is similar in kind to the naturally formed limestone video camera of Cave World, since we are able to gaze out upon Nature and glimpse the evolutionary forces that gave rise to us. Whereas the video camera of Cave World formed itself from the precisely orchestrated accretion of mineral limestone, the human nervous system and human cortex have formed themselves from the precise authoring and editing of DNA. In both cases, the force driving such organization rests with context, the context in each case being the prevailing laws of chemistry and physics. In Cave World, the natural laws of physics and chemistry serving as background context were such that the limestone naturally became deposited into the exact shape and structure of a fully functioning video camera. Similarly, in the real world, the various laws of physics and chemistry are such that matter naturally forms itself into molecular conglomerations that can evolve and eventually become so organized, so complex, that inquiring minds arise.

In considering the evolutionary emergence of the mindful human cortex, one simply must return to the question as to why Nature is this way. For if we saw some natural form like a stalactite exude itself into a highly intricate information-processing device able to learn about itself and examine the roots of its own creation, we would surely conclude that the natural laws that made this possible were reflective of an intention of some kind. Cave World would clearly be an intelligently configured world (just as artificial-life worlds are intelligently configured). The alternative, of course, would be to shrug nonchalantly and say that the natural laws of Cave World just happen to be able to do such things. Or maybe that there are an infinite number of differently configured parallel Cave Worlds.

Would you be happy with such a casual explanation? Of course not. So too then should we be dissatisfied with brute factual accounts as to why the real world is similarly blessed with ingenious powers of construction. That Nature was poised, from the very birth of the Universe, to eventually usher consciousness into existence via the medium of cortical bio-logic is remarkable, more so than the limestone video camera forming within Cave World. With hindsight, we *know for a fact* that the Universe, right from the word "bang," embodied the potential to eventually birth life and consciousness. And the only reasonable explanation for this is that Nature has some kind of inherent self-organizing, self-knowing function. *Thus the human cortex can be seen as an advanced means through which natural intelligence comes to reflect itself and know itself for what it is.*

The above statement is actually not quite as radical as it first appears. It is more of a factual statement at heart. Take the first part: *the human cortex is an advanced means . . .* Well, it is true by anyone's standards that the human cortex is an advanced piece of biological kit, certainly more advanced than any man-made machine. As for the rest of the sentence—*through which natural intelligence comes to reflect itself and know itself for what it is*—if we switch the term "natural intelligence" with "the Universe" or "the evolutionary process," then we again have a plain and simple statement of fact, much the same as the statement "the human cortex is the means through which the English language is developed" is a basic statement of fact. However, implicit in the statement is the saliency of the fact, for what the statement seems to be implying is that an essential function of consciousness is to allow natural intelligence to know itself. And to imply that is to imply that natural intelligence *intended* that the tree of life eventually produce conscious minds. To many, that will sound like heresy of the highest order.

It is interesting to see how some eminent scientists tackle this broad issue. Take the following quote from professor Christian de Duve, a Nobel Prize–winning biochemist who has written extensively on the inevitable formation of both life and conscious minds:

Trillions of biospheres coast through space on trillions of planets, channeling matter and energy into the creative fluxes of evolution. In whatever direction we turn our eyes when we look into the sky, there is life out there, somewhere. This fact completely alters the cosmological picture. The Earth is not a freak speck around a freak star in a freak galaxy, lost in an immense "unfeeling" whirlpool of stars and galaxies hurtling in time and space since the Big Bang. The Earth is part, together with trillions of other Earth-like bodies, of a cosmic cloud of "vital dust" that exists because the Universe is what it is. Avoiding any mention of design, we may, in a purely factual sense, state that the universe is constructed in such a way that this multitude of life-bearing planets was bound to arise.[1]

And further:

If the universe is not meaningless, what is its meaning? For me, this meaning is to be found in the structure of the universe, which happens to be such as to produce thought by way of life and mind. Thought, in turn, is a faculty whereby the universe can reflect upon itself, discover its own structure, and apprehend such immanent entities as truth, beauty, goodness, and love.[2]

Again, we must perforce ask ourselves why Nature "happens to be such." We have already ascertained that it is the contextual effect of the laws of Nature that, through their all-pervading sensibleness, serve to promote evolutionary processes. Laws engulf and saturate the Universe and foster all constructive processes, be they star formation or life formation or mind formation. This being so, it is valid to ask ourselves why Nature has these properties, why it is that so many fortuitous life-affirming circumstances abound. Why should Nature be such an inimitable life-support system able to elicit the genetic code and, later, conscious cortices? Why, indeed, should the Universe be endowed with such a tremendously creative potential? Christian de Duve seems, on the face of it, happy to shrug at Nature's specific configuration, as if it were of no importance (or maybe he is just being a trifle coy). But the facts cry out for interpretation. Is Nature intentional in some way?

THE HERESY OF TELEOLOGY

The thorny issue of teleology, which has here raised itself, must be dealt with. To contend that Nature is teleological is to say that Nature has an inherent purpose or an inherent function. It might also suggest that evolution to the point of consciousness had to happen according to some natural tendency of the Universe. Such an attitude is, perhaps, to recall religious paradigms, present in those oft-rosy belief systems that modern science has ostensibly banished to the compost heap of history, in favor of more empirically verifiable systems of belief. A cursory glance at popular science writing reveals that teleological speculation is anathema to most scientists, one of the cardinal sins against scientific reasoning. Teleologists don't get grants to do scientific research, since teleology has no place within the pursuits of merelyism. It is a pejorative word these days. If you confess to having teleological tendencies, you are doomed to ridicule and savage blows, it seems. Yet, as we have seen, natural intelligence is very much apparent in the configuration of Nature. In terms of the laws of Nature, these natural laws are such that matter is prone to self-organize and eventually morph into the ever-evolving biological systems of life, which learn to reflect, or mirror, this lawful sensibleness. Life can keep on learning only because there is so very much to learn about. Life can get smarter and smarter only because the larger system it is growing into is itself endlessly smart. And mind itself can arise and evolve only because there is more and more sense to be made of the Universe. The comprehensibility of Nature is profoundly extensive and is the source of most, if not all, cognitive nourishment.

In the ultimate analysis we saw that the only real way to deny natural intelligence and its contextual role in facilitating biological evolution is to appeal to some sort of infinitely bloated multi-Universe scenario in which everything and anything goes. Similarly, with Cave World, we can deny that its laws are deliberate and purposeful only if we invoke an infinity of Cave Worlds (each with different laws) and then conclude that we were merely imagining one of the more significant of them. If we grant that this line of thinking represents no more than a clumsy "dodge," which in the long run explains absolutely nothing (it just pushes the proverbial buck), then we are left with a view of Nature in which its natural laws must embody some kind of sensible design or self-design (as Darwin himself once remarked). And if this is indeed the case, then we have no choice but to entertain the radical idea that Nature is, at a fundamental level, teleological in some way, that Nature's unfolding constructive potential represents the ongoing realization of some sensible purpose. Not that this means that purpose was planted into Nature from *without*. On the contrary, it seems more reasonable to see such purpose as being part and parcel of a self-consistent Universe, a

Universe that, in its totality, is replete with self-organizing intelligence and whose meaning and raison d'être exists within itself.

At any rate, to suggest that natural intelligence is in the business of achieving some aim, that life and consciousness are *sense-making functions* of the Universe, is to explicitly imbue reality with a (natural) purpose. This is not necessarily a bad metaphysical stance when you think about it. Nor is it a crazy idea, for even if we were cajoled into imagining what a teleological Universe would be like, it is difficult to see how it would differ—in terms of its overall design attributes—from the one in which we now consciously exist (well, we might conceive of a "better" Universe, but it would be "better" only according to *our* standards). It is not like we are in poor metaphysical company here either, since all manner of esteemed scientific thinkers have divined intelligence and purpose within the Universe—geniuses of such caliber as Newton, Einstein, James Clerk Maxwell, and Kepler, for instance. As for the rather extraordinary consequences and implications of this view, we will consider these shortly. First let us return to the prime issue of contention at hand, namely the significance of human consciousness.

THE CENTRALITY OF MIND

The idea that consciousness, or mind, has a central role within the "intent" of Nature seems to place humans right in the fore of the action, as if the human race were the shining crown of creation. In other words, I could here be accused, among other travesties, of being overtly anthropocentric (i.e., of stressing man's importance). This is not strictly true however. What is important is not the human species per se but the sort of consciousness we *Homo sapiens* enjoy. It is precisely the conscious mind (as opposed to the hominid that happens to host it) that underlies science and the quest for the "truth" and is also able to comprehend the existence of natural intelligence. As stated in chapter 1, consciousness makes us what we are; it quite literally underscores our being. In this sense, consciousness can hardly be a mere aside to the ongoing drive of Nature to make sense of itself, and if I am to be charged guilty of anything, then it is in being "mindocentric." If a conscious mind, any self-reflective conscious mind in any kind of species, terrestrial or otherwise, can divine natural intelligence, this is surely suggestive of the significance of consciousness within the general scheme of Nature: a "mindocentric" belief perhaps, but certainly not anthropocentric.

Actually, the pursuit of merelyism, in which all vestiges of intelligence are remorselessly removed from life and its evolution, is far more guilty of being anthropocentric than is the belief in natural intelligence. This is because merelyism holds that only humans possess an advanced type of intelligence able to build and design complex things. Any scientist who denies that evolution bears the chief hallmarks of intelligence or denies that Nature is intelligently configured in terms of its laws is basically maintaining that human intelligence surpasses all other forms of intelligence—that, as far as we know, human intelligence alone is bound up with elaborately creative design pursuits. This is arguably a more unsound and conceited view than is a belief in natural intelligence and the significance of consciousness within its creative and purposeful schemes.

To boldly assert that consciousness is highly significant is indeed to place it at the fore of life (or the "leading edge" of evolving life). But, once again, it should be stressed that it is not the human race that is significant, rather it is the evolutionary genius inherent in the structure and functioning of the human cortex that is of most interest, for, as far as we know, without a highly evolved cortex there would be no (human) consciousness. The hard bone of contention therefore lies in the conjecture that conscious minds of one sort or another *had to evolve* since natural intelligence was, by default, determined to evolve them. If there were good grounds for believing such, then it would represent more evidence for the notion that Nature is bound up with an intentionality of some kind. But how on earth do we go about justifying the idea that consciousness of the sort we possess had to evolve somewhere along the line?

As difficult as it may seem, there is a way to explore this deep question. If it can be demonstrated that evolution must, by necessity, lead to the emergence of *specific* biological phenomena, then this would give added credence to the notion that natural intelligence, or Nature as a whole, is bound up with the realization of some sort of intent. This will be even more the case if those inevitable biological systems all have some property in common. The inevitable emergence of definite biological systems, one after another and all with some shared property, could then be seen as the progressive move toward a *target* inherent in the whole system of Nature. If it can indeed be shown that certain aspects of the tree of life had to arise due to some kind of natural logical necessity, then consciousness of the sort that the human brain conducts might well be one particular instance of such a natural necessity.

So what is the score on this issue? Does evolution move in *any* direction, or are there constraints of some kind? If evolution is all about the reiterative selection of sense-making systems of bio-logic, then is consciousness an example of a sense-making system substantiated by bio-logic and *bound* to emerge? To get to grips with this heady notion, we must start by considering attractors.

EVOLUTIONARY ATTRACTORS

Attractors, as the name suggests, are states within some system that the system inevitably steers itself toward according to the laws governing the system. In the case of a swinging pendulum encountering friction, the attractor is the state in which the pendulum sits motionless. No matter where you start the pendulum, it will always veer toward this attractor state; the law of gravity and the laws of thermodynamics ensure this. If you have seen the movie *Titanic,* the final state of the ship at the bottom of the ocean was an attractor, a state that, as the chief engineer chillingly remarks after the ship is struck, had, by dint of mathematical certainty, to be reached. Another example of an attractor is the state of match point in a game of tennis, since all games of tennis will, according to the specific rules of tennis, move inexorably toward that state. Rules are synonymous with laws—and rules/laws promote attractors.

Perhaps the most fundamental attractor within the evolving tree of life is its actual birth. We shall consider this before moving on to the idea that brain-embodied minds are likewise an attractor toward which evolving bio-logic inexorably gravitates.

As indicated by Duve, there is now a growing opinion among scientists that life, or replicating DNA, had to emerge because the very seeds of living processes are immanent within the cosmos. Given the sheer expressive power of DNA, it seems unreasonable to view the origin of life as stemming from a one-off freak accident. The same can be said of the origin of the genetic code. Or the origin of minds. Or even the origin of stable suns. It is all "far too good to be true." Life's evolution is better understood as a preordained unfolding potential "written into" the energetic fabric of Nature. Thus, life may, in itself, represent an initial attractor toward which systems of chemicals interacting with one another are inevitably drawn according to the directional dictates of naturally intelligent law. As Santa Fe complexity scientist Stuart Kauffman writes: "If we are . . . natural expressions of matter and energy coupled together in nonequilibrium systems, if life in its abundance were bound to arise, not as an incalculably improbable accident, but as an expected fulfillment of the natural order, then we truly are at home in the universe."[3]

The main thrust of Kauffman's widely cited work in this area revolves around an interesting model he has developed that explains how and why self-organizational systems of molecules spontaneously arise due to statistical probability. Once these systems have emerged, they can become subject to natural selection and, in consequence, become candidates for the origins of living things. It is worth looking at Kauffman's model in detail, since it furnishes our knowledge and understanding of natural intelligence in its more subtle aspects. The model also shows us that life represents an attractor, a state of self-organization that had to come to pass. Once we see that life itself is an attractor, we can go on to look at other natural necessities and judge whether conscious minds also fit the bill.

Kauffman asks us to imagine an initially lifeless soup of interacting polymers. Polymers are long-chained molecules that form themselves quite naturally due to the unique atomic properties of carbon, which allow carbon atoms to unite with other atoms in a long self-repeating sequence. In a soup, or puddle, or pond, of such polymeric molecules, some of them may react with one another so as to create *new* molecules. As the number of polymers increases, the number of possible reactions between them also increases.

The crucial thing to bear in mind here is that the number of possible reactions between the polymers increases *exponentially* as the number of individual types of polymers grows. You can understand this clearly if, instead of polymers, you think about individual letters of the alphabet and the number of combinations between them. If you have two letters, A and B, then there is only one reaction between them (AB). Introduce a third letter C and the number of possible reactions is now three (AB, AC, BC). Bring in a fourth letter D and the possible amount of reactions rises to six. A fifth letter E elevates the number of possible reactions to ten, the addition of a sixth letter F raises the reaction level to fifteen, a seventh letter to twenty-one, and so on. As the number of letters reaches double figures, the number of possible reactions escalates. Anyone who gambles will know of this exponential rise in combinations, or "doubles" as they are known in the betting world. It is the same with a lottery. The more numbers you pick from which unique combinations can be created, the more combinations there are to bet upon and the more costly the bet becomes.

Returning to the concept of interacting polymers within some primeval soup, Kauffman explains that if new molecules are constantly being created from various reactions between the polymers, soon the number of possible reactions within the soup becomes so high that some of these reactions will undoubtedly cause the formation of *already existing polymers.* In this way, *autocatalysis* may be born, a process in which an interacting set of molecules helps sustain itself by looping back on itself. Once more replacing polymers with letters, imagine that A and B react with one another to birth C. Let C react with A such that D is produced. If D then acts as a catalyst for the original reaction between A and B (i.e., D serves to promote A and B's interaction), then a catalytic set has emerged, which, by its very nature, will sustain its own cycles of creation. A kind of whole and stable system has suddenly "clicked" into place.

Here's Kauffman summarizing this compelling concept of autocatalysis and its role in the very beginnings of life:

> In this view of the origin of life, a critical diversity of molecules must be reached for the system to catch fire, for catalytic closure to be attained. A simple system with 10 polymers in it and a chance of catalysis of one in a million is just a set of dead molecules. Almost certainly, none of the 10 molecules catalyzes any of the possible reactions among the 10 molecules. Nothing happens in the inert soup save the very slow spontaneous chemical reactions. Increase the diversity and atomic complexity of the molecules, and more and more of the reactions among them become catalyzed by members of the system itself. As a threshold diversity is crossed, a giant web of catalyzed reactions crystallizes in a phase transition.[4]

A phase transition refers to a radical change in the material properties of a system. A drop of water freezing into a crystalline structure (i.e., the formation of an ice crystal or a snowflake) is a good example of a phase transition wherein water molecules suddenly assume a solid and robust patterned structure dramatically different from liquid water. In the case of Kauffman's envisaged autocatalytic set emerging from a polymeric soup, the pattern that suddenly "freezes out," or crystalizes itself, is a group of polymers whose reactions with one another cause their own formation. Autocatalysis is very reminiscent of autopoiesis, the process of self-generation we first met in chapter 2 that marks out living organisms.

The establishment of autocatalytic sets may well mark the humble beginnings of life. Since a self-sustaining network of polymers is not too different in principle from a self-*replicating* network of polymers (like DNA or RNA, which are further examples of polymers), Kauffman's idea is compelling. Moreover, autocatalysis results entirely from statistical probability married to the natural laws of molecular chemistry (these laws can, of course, be seen as natural intelligence in its most fundamental expression). In other words, given a vast soup of interacting polymeric molecules and given also precise chemical laws that govern their interaction, autocatalysis is *guaranteed to emerge at some stage*. And once it has emerged it could, in theory, provide the foundation for further "progress," at least in terms of the molecules becoming even more tightly bound to one another as well as becoming more efficient in promoting their own formation.

Kauffman has also noted other potential lifelike properties of an autocatalytic set. If the self-sustaining set grew—let's say it doubled in size—then it could conceivably break apart such that two autocatalytic sets become operational, in effect a process of simple replication. Autocatalytic sets can also be driven by sunlight, a source of free energy that could, in principle, help sustain, or feed, a nonequilibrium set of interacting and self-organizing molecules. Perhaps this goes some way in explaining why complex polymers, the seeds of life, have been found all over the Universe, including within meteors and comets.

It is worth mentioning here that the lifelike property of self-organization evinced by autocatalytic sets is also found in phospholipids. These too are naturally occurring polymers, in this case long chains of hydrocarbons (molecules of hydrogen and carbon) containing additional phosphorus and oxygen atoms. The phosphorous end of these phospholipids carries an electrical charge, which serves to attract water, whereas the other end of the phospholipid repels water due to an opposite charge. This kind of structural alignment of positive and negative charges is what causes oil to separate from water and thence form a thin film. In the same way, phospholipids in, say, the ocean will orient themselves with one face to the water and the other face away from the water. This process can lead to the formation of drops, or *enclosed spheres*. Double layers can likewise form due to the energetic action of ocean waves, thereby creating basic semi-permeable membranes alike in nature to those found in all living organisms. With a stretch of the imagination one can fairly easily envisage such spherical phospholipids reproducing themselves due to simple division, the effect of solar energy, or some other natural property of the world. Many experts believe life started in this way.

In both cases—Kauffman's autocatalytic sets and the properties of phospholipids—the process of key importance is self-organization. The laws of chemistry dictate that molecules react with one another, join with one another, relate to one another in precise ways, and so on, and this specific molecular behavior inevitably leads, on a more macroscopic level, to polymeric systems that have the inherent capacity to sustain themselves and even reproduce. Once this process has started, natural selection will have something to work upon, and differential reproduction can begin in earnest. Or at least that is the conclusion of this kind of research into self-organization. Kauffman argues that natural selection can work only hand in hand with self-organization and that until we acknowledge the spontaneous self-organizational properties of molecular matter, we cannot fully invoke the principle of natural selection. Only when self-organization of the sort Kauffman attests to has transpired can evolution through natural selection take hold. Which is to say that molecular self-organization provides the initial complex systems that natural selection can then set to work upon. The seed of our tree of life was thus likely composed of "primitive" self-organized matter.

Back to attractors. If Kauffman is correct and autocatalytic sets of interactive polymers are guaranteed to emerge according to natural law, then these kinds of self-sustaining systems can be viewed as attractors, material systems that simply had to manifest somewhere and somewhen within the Universe. More to the point, the only reason a catalytic set can arise is courtesy of the contextual effect of the laws of physics and chemistry, which ensure that atoms form into molecules and that molecules form into macro-conglomerations of polymers with potentially significant relations between one another. Through energetic affinity, matter binds itself and weaves itself into patterned arrangements, these arrangements made sensible through the equally sensible effect of natural laws. Life emerges not as a strange and bizarre anomaly but as an attractor toward which molecular systems are naturally driven. Given the right conditions, planets can "catch life" just as wood can catch fire.

CONVERGENT EVOLUTION

More obvious evidence for the operation of attractors within the web of life stems from a process known as convergent evolution. With evolutionary convergence, a basically identical biological or behavioral system manifests within widely different species of organism. The two most well-known examples of convergent evolution involve eyes and wings. The eye, for instance, has evolved more than twenty times in separate branches of the tree of life. Similarly, wings developed independently in insects, birds, and bats. Other examples of convergent evolution include echolocation, which evolved independently in whales and bats. Likewise, the similar fusiform shape of dolphins (mammal) and sharks (fish) highlights an "attractive" solution to the problem of how to move fast in water. The symbiotic alliance between ants and fungi detailed in the previous chapter is echoed among some species of termite, and this is another case of convergent evolution. And if you recall the microscopic branching pattern of moth antennae, this kind of fractal structure is also found in other insects like the praying mantis, which is an insect belonging to a totally different order (organs like lungs also branch in this fractal way). Even the kind of remarkable endosymbiotic breakthroughs detailed in the preceding chapter are thought to have occurred more than once. Algae, for example, are believed to have evolved symbiotic relationships with bacteria on at least three independent occasions.

In the case of the convergent evolution of eyes, it is the case that light is itself an eminently sensible aspect of Nature. Unimaginably rich in information, light was, as with other manifestations of Nature, always waiting to be made sense of even before eyes evolved to become sensitive to it. If you have to live and reproduce in the world, it makes great sense to model the world, to "see" the world in whatever way you can. Once an organism can register light, no matter how feebly, the organism is in a better position to survive and reproduce. Seeing by making use of light through systems of bio-logic that are sensitive to light therefore represents a biological attractor, and this adequately explains why visual systems have evolved independently in so many branches of the tree of life.

In each case (and there are lots of other examples), evolution arrives at similar solutions and similar patterns of biological design. If we embrace the paradigm of natural intelligence, then it is quite obvious that in the ongoing drive to make sense of the environment and learn more about it, bio-logic is obliged to evolve in certain specific directions according to the already existing specific sensibleness that it is exposed to. The independent evolution of complex eyes in multiple branches of the tree of life ably highlights how different systems of bio-logic will be honed by Nature into embodying the same kind of form and function. Sensible solutions to the art of living are therefore attractors that the tree of life is obliged to gravitate toward.

CONSCIOUS BRAINS AS ATTRACTORS

The question we have been steadfastly approaching concerns the emergence of conscious minds, the type of mind that we humans have and that can begin appreciating natural intelligence. Was it inevitable? In terms of sense-making, is mind a kind of attractor toward which biological evolution will inevitably be drawn in the same way that eyes, wings, and body shape are inevitable biological attractors? If, on the other hand, mind is not a "cortical attractor," then it implies that even if natural intelligence were real, and even if natural intelligence were indicative of a manifesting intent of some kind, a drive to evolve conscious minds might not be part of Nature's chief agenda. Although it might be intuitively difficult to reconcile such reasoning with the paradigm of natural intelligence, we must consider the evidence at hand before drawing any conclusion.

While this is certainly an unusual area of inquiry, it has actually been dealt with at length by evolutionary biologist Stephen Jay Gould in his classic book *Wonderful Life*. Although Gould takes the reader on a lengthy preamble, the gist of the book concerns itself with the question of whether humans (and by extension conscious minds) are an inevitable outcome of evolution or whether the human race is simply the result of happenstance.

Gould initially takes his readers on a tour of the Burgess Shale, a famous geological site in the Canadian Rockies awash with decidedly strange fossils dating from the Cambrian explosion some half a billion years ago (it is referred to as the Cambrian explosion because a plethora of new life-forms sprang from this era). The uncanny thing about some of these fossils is that they represent organisms so different, so weird, that they have been classed as belonging to no-longer-existing *phyla*.

I use italics for good reason. Indeed, here lies the principal reason for Gould's descriptions of the Burgess Shale. If you consider the fact that the five kingdoms of life (animals, plants, fungi, protoctists, and bacteria) are initially divided into various phyla and that phyla are further divided into classes and these classes further subdivided into orders and so on all the way down to individual species, then fossil evidence indicative of an extinct phylum represents a major find. It implies that had the phylum not gone extinct, life on Earth might have been very different from what it is now. This is because a phylum is such a large grouping. If a single species of organism had not gone extinct in the distant past—and bear in mind that most species *do* go extinct—we might not expect too much difference in the overall fate, or shape, of the tree of life thereafter. Even if one of the many no-longer-existing genera had not vanished into oblivion but had survived, we should still not expect things to be that altered. But if an entirely unique phylum had survived instead of perishing, we should expect the tree of life to be very different from what it is now.

We and other animals like fish, birds, amphibians, and reptiles belong to the chordate phylum, the principle characteristic of which is the possession of a backbone. And fossils of the oldest known chordate have actually been found in the Burgess Shale. Amazingly, the fossilized remains of *Pikaia,* a wormlike creature found in the Burgess Shale, is believed to represent the ultimate ancestor of all of today's living chordates, including us. But had *Pikaia* gone extinct and had a member of one of the other of Burgess's peculiar phyla succeeded instead, then chordates like birds, fish, and mammals might not be alive today but might have been replaced with altogether different animals.

As to why *Pikaia* survived and went on to spawn all of today's mammals, fish, and birds, Gould believed chance played a key part. Since I am not an expert on fossils and extinction events, we will take his word for it and accept that the reason *Pikaia* survived and other creatures did not was due more to chance environmental events than to specific biological design (then again, maybe *Pikaia* embodied a more sensible form of bio-logic). Thus, Gould's studies of the Burgess Shale led him to conclude that human life—a distant offshoot of the chordate phylum heralded by *Pikaia*—is an entirely chancy and improbable event, and that if you were to run evolution again any number of times from the Cambrian explosion, the human species (and human history) would be unlikely to emerge:

> Little quirks at the outset, occurring for no particular reason, unleash cascades of consequences that make a particular future seem inevitable in retrospect. But the slightest early nudge contacts

a different groove, and history veers into another plausible channel, diverging continually from its original pathway.[5]

The simple way to think about this is to imagine that the evolution of the tree of life is an old-fashioned videotape, which can be rewound and restarted. Rather than the tape being fixed, we can imagine that each time we replay events from the Cambrian age five hundred million years ago, a different tree of life emerges. Every time we replay the tape we witness a different evolutionary tale unfold. This implies that evolution produces much that is contingent—which means that these things depend upon complicated webs of chance. What is not chancy or contingent, of course, are the general laws of evolution, which ensure the emergence of good design and good behavioral strategies.

The lesson to be drawn from the Burgess Shale is humbling. Gould's basic premise is that *we did not have to be here,* that the only reason we are here depends upon chance and not necessity. As he states:

> And so, ultimately, the question of questions boils down to the placement of the boundary between predictability under invariant law [necessary forms of biology] and the multifarious possibilities of historical contingency . . . I envision a boundary sitting so high that almost every interesting event of life's history falls into the realm of contingency. I regard the new interpretation of the Burgess Shale as nature's finest argument for placing the boundary this high.[6]

If we grant that the evolution of humans is an "interesting event," then Gould is maintaining that this state of affairs is purely dependent upon chance. Run evolution again and the odds are that humans would not evolve. Perhaps *Pikaia* would have become extinct, thus stopping the subsequent development of mammalian chordates like us. Alter some biological eventuality half a billion years ago and, even if the alteration was relatively minor, the fate of the tree of life changes dramatically. Life maybe inevitable, but its actual shape, the way in which it actually manifests, depends upon chance rather than necessity. Kauffman's earlier testimony could therefore be amended so as to suggest that life is certainly at home in the Universe but that *we* are nonetheless lucky to be here. Or so Gould would have us believe.

So where does all this reasoning leave us? As mindful hominids scratching our well-endowed craniums over the sheer odds against our being here? I have already stated that, with regard to consciousness, the fact that the human brain happens to be conducive to conscious processes is not really the point. The possible contender for an attractor was consciousness itself and not the type of species that is endowed with it. This point of view serves to dampen Gould's otherwise deflationary remarks. At any rate, coming back to the boundary Gould spoke of, the boundary between predictability and chance, Gould candidly states: "Whether the evolutionary origin of self-conscious intelligence in any form lies above or below the boundary, I simply do not know."[7]

On the face of it, Gould seems to have an open mind on the issue, allowing that it remains a possibility that even if the present human race is dependent upon chance evolutionary events, the existence of advanced consciousness in some species or another might still be an inevitable outcome of all possible runs of the evolutionary tape. However, Gould later writes:

Since human intelligence arose just a geological second ago, we face the stunning fact that the evolution of self-consciousness required about half of the earth's potential time. Given the errors and uncertainties, the variations of rates and pathways in other runs of the tape, what possible confidence can we have in the eventual origin of our distinctive mental abilities. Run the tape again, and even if the same general pathways emerge, it might take twenty billion years to reach self-consciousness this time—except that the earth would be incinerated billions of years before.[8]

To counter, it must be said that even if the Earth had not had time enough to evolve a conscious brain, we can bet that *another* biosphere would have had plenty of time to do so. Or even if humans had not evolved here on Earth, maybe another similarly conscious species would have evolved instead. Even if belligerent dinosaurs still ruled the Earth and were somehow incapable of cortical evolution to the point of refined conscious intelligence, this does not preclude a scenario in which mammals of some kind evolve self-reflective consciousness. Incredulity should not seal our minds here; rather a use of our imagination can yield hypothetical creatures who, like us, can do science and art and learn of natural intelligence. And this is precisely the point. It is highly likely that somewhere in the Universe, within some biosphere, a nervous system, at some time, had to evolve the capacity to embody consciousness of the sort we have. It just so happens that this has, at the very least, transpired here on Earth in this part of the endlessly expansive cosmos. If life really is inevitable, as Stuart Kauffman, Duve, and many others suggest, if it is wholly natural for planets to "catch life" when certain informative conditions prevail, then there will almost certainly be countless biospheres throughout the vastness of the Universe. And if biological evolution—by way of DNA or some other natural polymeric language—is likewise inevitable, then it is reasonable to assume that mindful brains represent a further inevitable consequence of Nature's evolutionary craft.

NERVOUS SYSTEMS AND CONSCIOUSNESS

Needless to say, it is nigh on impossible to conclusively *prove* that self-reflective consciousness has to emerge in any particular tree of life. But, as alluded, there is much to support the contention. If evolution is viewed as a sense-making process, then nervous systems—part of which is the brain—are obviously synonymous with sense-making systems. That's exactly what they are, and that's exactly what they do. Composed of networks of excitable neurons or nerve cells, nervous systems access, store, and organize large amounts of information about the world. This information, arriving through different sense organs like the eye, the ear, the skin, and so forth, is then organized and integrated by some clever means in the brain, which can be considered the "highest" part of a nervous system. There will clearly be a fitness premium on those nervous systems that can maximize this kind of information processing. Hence we find the complexification of nervous systems and brains over large spans of evolutionary time. Visual systems become more complex, auditory and olfactory systems become more acute, and so on. All animals have nervous systems of one sort or another. And even the bizarre creatures of the extinct phylum of the Burgess Shale had nervous systems. They were, after all, still bona fide members of the animal kingdom. So once a nervous system has evolved in an organism, natural evolutionary logic dictates that there will be a selective pressure for these nervous systems to become more adapted, more efficient in making and collating sense. Hence, brains get bigger, smarter, and more sensitive. This is essentially a biological attractor, or a *direction* in which natural intelligence will steer biology. As Duve suggests:

> The direction leading to polyneuronal circuit formation [nervous systems] is likely to be specially privileged . . . so great are the advantages linked with it. Let something like a neuron once emerge, and neuronal networks of increasing complexity are almost bound to arise. The drive toward larger brains and, therefore, toward more consciousness, intelligence, and communication ability dominates the animal limb of the tree of life on Earth, and could well do so on many other life-bearing planets. On the other hand, the bodies serving the brains and controlled by them need not be similar to human bodies, although they would likely possess appropriate means for sensing, acting, and communicating.[9]

Duve's point is well put and well taken. Polyneuronal circuit formations—for example, nervous systems and brains—*are* biological attractors, just as eyes, legs, wings, metabolic pathways, symbiotic unions, and semi-permeable membranes are biological attractors. Indeed, this explains the exceptionally rapid evolution of the hominid brain over the last few million years. The only good reason for this is that bigger brains allow more sense to be made of the world. Once more sense is made, more understanding and ingenuity can be applied to the arduous task of living and surviving in the world. Even the unique language faculty of the human brain pays testimony to sense-making, since the use of language allows sense to be conveyed and stored. Brains and nervous systems quite literally construct good sense (cognitively speaking), and this is why their evolutionary emergence is assured.

So no matter how many times you rerun Gould's tape of life, certain biological manifestations will always resolve themselves in the same way that Dawkins's fish eye algorithm will always resolve a fish eye no matter how many times the algorithm is run. What drives the emergence of conscious brains (and informs them) is the comprehensibility and sensibleness of Nature, neuronal brains being a particularly refined biotechnological means with which to reflect or mirror that sensibleness. The more evolved the brain (or polyneuronal network), the better able it is to make sense of the contextual environment of Nature in which it finds itself, a contextual environment that, through its specific configuration, was always waiting to be made sense of. And so here we are, attached so to speak to a hominid body, which, unlike consciousness, *could* have been otherwise. While the human species might be completely beside the point, the sort of consciousness our brains convey simply had to arise somewhere along the evolutionary line.

NECESSARY STEPS

Interestingly, Gould goes on to question many other things that seem to be special about life. This is really another form of merelyism in which the most spectacular feats of bio-logic are considered the result of chance events as opposed to the necessary expressions of biologically embodied intelligence. For instance, Gould talks about the ostensibly chancy nature of symbiosis wherein prokaryotic cells (cells without a nucleus) evolved, through symbiosis, into eukaryotic cells (cells with a nucleus) and thence eventually into integrated multicellular organisms. Gould accepts Lynn Margulis's theory of endosymbiosis, which we encountered in the preceding chapter, and which we rightly marveled over. However, Gould asserts that this process was the result of contingency rather than being lawfully bound to happen. He writes:

> If you wish . . . to view the origin of organelles and the transition from symbiosis to integration as predictable in some orderly fashion, then tell me why more than half the history of life passed before the process got started.[10]

But time is surely of no consequence. Time is a wholly relative concept. It makes no difference whether a crucial stage in the growth of the tree of life took billions of years or millions. You could equally ask why life on Earth started three and a half billion years ago and not a couple of million years earlier or later. Indeed, if life *could* have started a million years earlier than it did, would that imply that Nature is a bit slack or sluggish? What really matters of course is the resultant phenomenon in question—bacterial endosymbiosis—and whether such an eventuality was based upon chance or upon necessity. As it is, endosymbiosis allowed multicellular organisms and nervous systems to flourish. Since, with hindsight, we can see that tight symbiosis—such as those instances involving mitochondria and chloroplasts—allows individual embodiments of intelligence to coalesce and thus make more sense together than apart, it is evident that this is not a chance event but rather an inevitability, one of the many logical outcomes of natural intelligence as it expresses itself in the self-assembling world that is Nature.

No doubt about it, endosymbiosis represents one particularly shrewd solution on biology's road to making ever more sense. Or, if not a solution, then at least a technique. Once this skillful step was taken, once this technique was arrived at, life never looked back, and the plant and animal kingdoms bloomed, eventually going on to produce conscious nervous systems, themselves orchestrated out of endosymbiotic wizardry. It does not matter how long the first instance of an endosymbiotic solution took in coming. Nor does it matter why it took billions of years for bacteria to join together in a lasting symbiotic embrace. This is basically the same time-dependent principle of self-organization we met with earlier in Kauffman's account of autocatalytic polymer networks. Given enough time (and Nature has all the time in the world) and given also the sensible context of Nature, astonishing sense-making processes such as autocatalysis and endosymbiotic orchestration will occur. These events had to happen somewhere, at least on one biosphere, if not in each and every one of them. And we, at least, are here to prove it. In the last analysis, our conscious existence pays testimony to the fact that if it can happen once, it can happen again. Endosymbiosis, as with consciousness, can thus be placed in the category of necessity as opposed to mere chance. In short, anything that *can* fit together and make sense *will* eventually fit together and make sense. Or, to put it another way, if information can fall into place and is encouraged by law to fall into place, then it *will* fall into place. Such is the self-assembly nature of the cosmic jigsaw.

What would genuinely be chance, of course, is the actual moment and the actual location of the first endosymbiotic step, as well, perhaps, as the exact kinds of microorganisms involved in the neat equation. So too might we view much of the subsequent action—the elaboration and evolution of those first endosymbiotic organisms—as governed by contingency. Which means that many of the directions toward which evolution gravitates are determined (or intended, as the case may be), whereas the precise details are left to chance. Just as an apple tree will *have* to grow a trunk, branches, leaves, and roots, all with a definite form or structure, the actual *precise shape* of those organs will depend upon webs of contingency. As Darwin intimated, designed laws will manifest something definite, whereas chance will be evident in the details.

CONSCIOUSNESS

A Fulfillment of the Natural Order

The message of this chapter can be succinctly stated as follows: It is not that *we* are at home in the Universe, but rather that *our consciousness is*. Or, to put it another way, only if we think of "we" as relating to conscious experience as opposed to the human organism can we rightly claim to have been expected all along by Nature. Each of the most crucial of the steps that got us here—that of life's initial emergence from the primeval soup to endosymbiotic breakthroughs, multi-cellularity, and the subsequent emergence of both nervous systems and conscious brains—can be seen as progressively inevitable phases in the eminently creative schemes of natural intelligence. With evolution, some things are guaranteed. That which makes sense both in and of the contextual light of Nature is that which will come to be, that which will be fulfilled. While the thick, leafy canopy of the tree of life might be the result of contingency, the essential underlying form of the tree is nevertheless due to necessity.

Conscious intelligence, being a sense-making technique or sense-making solution, had to be; it had to be facilitated into being via the biological evolution of polyneuronal circuitry. We know that it happened here on Earth. The chances are that in other trees of life in other parts of the Universe, self-reflective consciousness of our kind similarly exists, even though the organism so endowed might, on the face of it, be wildly different from the human organism. What will be similar will be an underlying nervous system of some kind capable of accessing and organizing information to the extent that conscious intelligence emerges. All that really matters at this point is the striking idea that Nature is verily determined to cultivate consciousness, that minds capable of grasping and contemplating natural intelligence were destined to emerge somewhere and somewhen within the tree of life. Heretical or not, let us pursue this extraordinary line of inquiry.

CHAPTER 10

THE CONSCIOUS
TREE OF LIFE

Through the inevitable evolution of nervous systems, or polyneuronal circuitry as Duve terms it, the tree of life has now become conscious of itself. Evolution has become conscious of itself. Indeed, natural intelligence has become conscious of itself. Or perhaps it would be more accurate to say that natural intelligence is *becoming* conscious of itself. In any case, it is consciousness that can grasp natural intelligence, feel it, and even intimate that such an experience is a means through which Nature can appreciate itself and know itself for what it is. The possibilities and implications of this admittedly poetical idea seem, at least to my mind, to be fascinating and endless, reminiscent of the process of awakening, as if the biosphere (and beyond) were quite literally waking up. But before we explore this idea, let's take one last look at the role of context in determining the tree of life's emergence as well as the shape of its growth and development. Any remaining difficulties in conceiving of natural intelligence in its more subtle holistic form will hopefully be vanquished by what follows.

It is customary to think of the tree of life as a real sort of tree with a trunk, branches, twigs, and so forth. As a descriptive tool, the use of a common tree to symbolize the long evolution of life allows us to see all of evolution at once and in a way that conveys fluid growth and progressive development. This explains the popularity of such an image. Of course, to work properly upon the mind, the archetypal tree of life has to be taken as a four-dimensional structure capturing the passage of billions of years worth of time. Which means that the main base of the tree of life, its trunk, represents primeval life, that is, single-celled bacteria—the monera kingdom, the first form of life. The other four kingdoms—plants, protoctists (sometimes called protists), fungi, and animals—are represented as four large branches that sprout from the main trunk. The topmost part of the tree (or, alternatively, its outermost surface) therefore represents the most recently evolved creatures. In simple terms, the tree's bushy foliage represents all extant species.

If we imagine the tree of life sprouting on a computer monitor, we can probably picture it growing before our eyes in the same way that a normal tree would grow, only in this case the pixelated cells of the tree represent actual species of organism. Cells, or blocks of pixels, building upon one another, represent new species evolving from already established species.

As we have seen, although the actual detailed shape of the tree of life may be dependent upon chance factors, the underlying shape is nevertheless determined by natural intelligence in its primary form, which, as the reader will recall, we previously referred to as being the various laws of Nature, laden as they are with sensibleness and the contextual ability to drive law and order. The presence of such a lawful and sensible context ensures that certain *specific* biological structures and behaviors will arise because of the inherent constraints in the biological art of sense-making. Indeed, it should be crystal clear by now that were it not for the contextual effects of Nature's laws, there would be no tree of life at all. Sensible and lawfully ordered contexts are always required to imbue structures—in this case evolved biological structures coded for by DNA—with sense and meaning. So what we really need to do with our tree image is to somehow visibly expose the contextual presence of the rest of Nature, which evokes and provokes the tree's growth. In other words, we need to be able to visualize the lawful presence of natural intelligence (i.e., its primary form), which the tree of life, in effect, reflects. Which is why another representation of the tree of life is useful here—not a new one, but a more concise one in which we can more properly appraise natural intelligence in its primary contextual role.

Imagine a computer screen before you and the tree of life in white set against a black background (as in a simple 2-D animation). The tree starts off as a small white structure, which gradually grows upward and outward, branches forming and further dividing and so forth. We can even envisage that the hidden roots of the tree are bound up with basic processes of autocatalytic self-organization.

At first glance we think that the white tree is simply growing into a black void, and that it could grow in any direction. But we know that if we posit the existence of attractors, the tree cannot grow in just any old direction or according to any old shape but must assume certain specific "postures." These necessary postures correspond to biological attractors, sense-making systems of bio-logic attuned to the laws of Nature and the sensibleness of the environment. Thus, in order to fully realize the fundamental contextual effect of Nature, we must strive to envisage that natural intelligence is initially represented by the black background into which the tree grows. However, rather than being an insubstantial featureless void, the black background actually serves to both draw out and highlight the tree, as if the white tree were a flexible substance being cajoled into a mold. This implies that *the black background of the white tree is just as much a part of the tree as the tree itself,* and that the tree molds itself according to the black background.

This, of course, is precisely what context is all about. Context *defines* things and gives things their meaning. A white tree on a computer monitor stands out only because of the black context surrounding it. This black contextual backdrop highlighting the white tree is synonymous with natural intelligence in its primary form as the various laws of Nature. Although this may still be hard to envisage (we find it easier to see familiar shapes than we do the backgrounds surrounding those shapes), if you can manage to give equal conceptual importance to the black background surrounding the white tree, then you will perceive natural intelligence in its primary form on a kind of visually emotive level. It takes a little effort, but a switch in perspective can be readily achieved.

As far as we know, the tree of life reaches its most intimate relationship with its background with the emergence of mind, for it is mind that can behold, at least potentially, the full extent of the naturally intelligent context in which it has arisen. *Mind is thus akin to a mirrorlike surface adorning the outermost layer of the tree of life.* Conscious minds are no less than localized reflections, or acute reverberations, of that which animates and nourishes mind. And so we have arrived back at the amazing phenomenon of consciousness, that very aspect of the tree of life whose import we readily gleaned in the opening chapter of this book.

AN ULTIMATE ATTRACTOR?

With the attractor-like emergence of mind ushered in through the evolution of the human cortex, sense-making—the hallmark of all living systems—has assuredly moved to a much more refined level. Sense-making now becomes a *moment-by-moment* enterprise. The sense-making capacities afforded by conscious intelligence are immense. Being plastic and highly adaptive, the conscious mind can alter its structure so to speak in order to accommodate more and more information about the world and thereby continually make more and more sense. Whereas most biological structures are basically (but not absolutely) fixed in place—think of cells, bones, internal organs, and so forth—consciousness is not quite so fixed. Bio-logic may be flexible, but mind is *ultra*-flexible. So whereas bio-logic can evolve, morph, and become more accurate in its sense-making capacity only over millions of years, the mind can adapt and "evolve" in far shorter spans of time. Mind thus learns and "understands" at a more rapid speed than biology.

Our feet, for example, are pretty much fixed in shape and are designed for walking on essentially soft, flat, dry ground (with maybe a bit of climbing and paddling thrown in). Minds are different. Minds can adapt to wildly different environments almost instantaneously. So even if the ground is rocky, jagged, or frozen and not as comfortable as the dry, grassy plains upon which our distant African ancestors may once have roamed, mind can make sense of harsh terrain by designing protective, malleable footwear. Similarly, if the environment suddenly becomes dangerously cold, instead of waiting countless generations to evolve thick fur or thick skin, conscious intelligence can "immediately" design warm clothing or learn to make shelter. So we can say that brain-embodied minds represent mutable and adaptable sense-making systems on a more advanced level than biology, even though it is biology that has facilitated mind. Or, if not more advanced, then mind at least represents a *new dimension* in the art of sense-making. The great flexibility of mind, its remarkable capacity to know, learn, understand, and make sense of the world, is precisely its chief virtue in terms of evolutionary viability.

Many are the ways in which the conscious mind and its sense-making potential can aid the organism with which it is associated. Think of the treatment of a wound through the deliberate application of some traditional herbal ointment, or the building of a water mill, or the use of language to convey how to make a fire. In each case, the ingenuity of the conscious mind can help elevate the fitness of the entire organism (or social group). Mind, being able to make more and more sense of the environment and being able to convey this sense through the medium of a language-using culture, is therefore a highly profound evolutionary development, just as extraordinary as the evolution of life onto the land and other epic moments in the growth of the tree of life. Indeed, one has only to stand back and contemplate the strange explosive drama of human history to realize the power and sheer biospherical impact of the conscious human mind. There can be little doubt that consciousness augments and enhances biological survival.

SENSE-MAKING, SENSE-SEEKING MINDS

The sense-making and sense-seeking properties of mind are powerfully expressed throughout the long sweep of human history, especially in the birth and growth of the sciences. Human culture's oldest science—astronomy—for example is concerned with the lawful movements of the heavens. Calendar systems, temples built to map the stars, knowledge of the procession of the equinoxes, myths associated with constellations—those are all cultural reflections of the non-random movement of the enduring stars above us.

The reason that the heavens move in a sensible and predictable fashion once again lies with natural intelligence in its primary form, namely the laws of Nature, which, in the case of the law of gravity, dictate the orderly relations between heavenly bodies. Indeed, the non-random movement of the stars provides us with a particularly vivid example of the secondary effects of natural intelligence, which arise according to its primary lawful form. For without the primary laws of Nature—such as gravity—there would be no order whatsoever. And if there were no order there would be no sensibleness and therefore no evolution of any kind. Laws, like the law of gravity, broadcast their sense throughout the Universe. And it is mind that, like bio-logic, learns to reflect, or make sense of, the various effects of natural laws. Whereas bio-logic can manipulate and make sense of molecular information, mind manipulates and makes sense of more widespread worldly information. The principle of sense-making remains the same. One can even think of "organs of thought" or "tissues of thought" (i.e., conscious informational structures), or *memes,* which, like biological organs, are sensible and can be further evolved and honed within a cultural, as opposed to a biological, milieu.

In speaking of gravity, although it was Sir Isaac Newton who first formulated the laws of gravitational motion in the seventeenth century by giving them a precise mathematical description (the famous inverse square law), the orderly movements of the stars were plain to see for our prehistoric ancestors. No doubt the consistent patterns within the starry sky (which would have been much more visible and breathtaking in times past due to an unpolluted atmosphere) filled our ancestors with awe and made them feel intimately tied to some larger cosmic system. Perceptions of this sort—like perceiving the ordered movement of the constellations, for example—most likely served to fuel the human mind, stimulating it to become more and more open in its receptive capacity and its learning potential. At any rate, the grand order and grand intelligibility inherent in the night sky explains why, for many ancient cultures, astronomy was one of the most valued sciences, and why early religion was intimately bound up with the non-arbitrary movements of the stars. Intimating the orderly behavior and ultimate coherency of the heavens instantly integrates us into a vast and meaningful context (it is food for thought to realize that an unobscured view of the starry night sky, while having no obvious economic value, might nonetheless be *invaluable*).

In pursuing the science of astronomy, then, human consciousness was becoming wise to certain orderly and sensible features of natural intelligence (i.e., that of the widespread effects of its primary form). Just as natural intelligence in its primary form as the laws of Nature serves to encourage secondary expressions like the non-random formation and non-random movement of galaxies, stars, planets, and evolving trees of life, so too does this informative influence find its way into the collective human mind. One can well imagine the first human or proto-human who noticed the seasonal movement of, say, the Plough, which consists of the seven brightest stars in the constellation Ursa Major (it is my favorite constellation and I call it the "Saucepan"). Although the actual configuration of the seven stars of the Plough would have appeared slightly different hundreds of thousands of years ago, their pronounced arrangement and seasonal movement would not have been lost to the hominid mind for long. Similarly, bright planets like Venus and Mars would have sparked continued interest, particularly with regard to their erratic seasonal movements (they behave like "rogue stars"). Likewise with the waxing and waning of the moon. Eventually the human mind began to sense that some sort of order and coherency governed the motion of heavenly bodies. This is a fine example wherein natural intelligence, in terms of its primary lawful aspect, gradually reflects itself in the realm of conscious awareness via the perceivable effects of gravity upon huge conglomerations of cosmic matter. Humble beginnings perhaps, but it was surely upon such "nourishment" that the human mind fed, grew, and expanded.

Apart from the distant heavens, the conscious human mind has also learned to make sense of the apparent movement of the star closest to us—namely the sun. Just as the compass termites we met in chapter 7 have learned to make sense of the sun's orderly movements relative to them, so too have human minds. The dynamics of the Earth's relationship with the sun, the inexorable flow of the seasons and such, all depend upon the effects of natural law, which governs, or orders, the precise movement of the Earth around the sun. In the same way that organisms like plants can evolve beautiful biological rhythms that resonate with the passage of seasons, so too can the conscious mind learn about seasons and anticipate them. The planting of crops at definite periods of the year or the gathering of fuel for winter once again demonstrates the capacity of consciousness to reflect law-conforming environmental events. Nature's law and order thus furnishes and stimulates conscious intelligence. Human progress is always bound up with the evolution of knowledge and understanding, a process that, like biological evolution, is driven by the tutorial influence of Nature. The order of the cosmos plants order in the mind. The whole informs the parts.

In one way or another, then, the widespread effects and properties of natural intelligence were bound to introduce themselves to the human mind. And different aspects became known at different times. For all his philosophical wisdom, Aristotle knew not of the endosymbiosis transpiring within him. Newton knew not of the genes and DNA underlying his keen sense organs. Darwin knew not of supernova events and how they generate the elements of which he and the rest of organic life were/are woven. Einstein knew not of the quarks dancing to some strange quantum beat within the atoms of his body. Natural intelligence in one form or another has constantly been making itself known to consciousness. My suggestion is that once upon a distant time the ordered movements of the sun, the moon, and the stars first galvanized the human mind into forging a connection with the larger prevailing system of natural intelligence. Maybe astronomical insights of this sparked the very beginnings of enhanced conscious life.

SCIENCE AND SENSE-MAKING

With the continual progress of science—a classic sense-making enterprise—the human mind has become more and more accurate in its reflection of natural intelligence, more and more adept at comprehending the various manifestations of Nature. The sciences of biology, biochemistry, genetics, chemistry, and physics, for instance, all tap into particular aspects of natural intelligence. Biochemists and geneticists focus upon biological reflections of natural intelligence. They tease apart and "reverse-engineer" bio-logic. Chemists concentrate upon the sensible behavior and the orderly relationships of elements with one another, these relationships again stemming from the effects of natural intelligence in its more primary lawful form. Physicists likewise base their often esoteric knowledge upon natural intelligence in its primary guise, by studying the lawful ways in which matter and energy behave. Almost all science—its name meaning "knowledge"—is based upon the study of natural intelligence in one form or another.

Once Nature has become understood by conscious minds, knowledge can be put to practical use (for good and for ill). In the same way that biological evolution is tuned in to an intelligently configured environment, so too is science and technological innovation tuned in to the same rich source. Take our discovery of electricity. The electrical properties of atomic matter are in the nature of things and existed even before science discovered these properties and first learned how to harness them. In fact, as with many of our inventions, biology got there first. The neurons currently firing in your brain make use of electrochemical potentials, that is, the electrical properties of molecular matter are exploited by nervous systems in order to transmit information. The use of electricity by our culture is a classic example in which the human mind has, like bio-logic, made fruitful sense of this potent electrical aspect of matter. The same holds true for most, if not all, of our scientific inventions such as iron smelting, nuclear fission, and genetic engineering. Since mind is inextricably welded to the sensibleness of Nature, mind can learn about this sensibleness and consequently build upon it. It appears that the only real difference between bio-logic and consciousness is the *rate* and *extent* to which natural intelligence becomes reflected.

In instigating a thorough dialogue with Nature, scientists have thus become privy to the wiles and ways of natural intelligence. Returning to the idea that the tree of life is a reflection of natural intelligence, the sum total of scientific knowledge about Nature must represent a particularly accurate and intimate component of this reflection. Given that science has learned the major laws of Nature and given also that the laws of Nature are synonymous with natural intelligence in its primary form, then scientific knowledge really is a reflection of natural intelligence. Our scientific understanding of Nature emerges as a conscious outgrowth within the otherwise physical tree of life, moreover an outgrowth whose function would appear to be bound up with the manifest inclinations of natural intelligence. In other words, this conscious outgrowth of which we are a part can, perhaps, be seen as a kind of tool, or technique, through which a high-resolution reflection of natural intelligence is more fully effected. Indeed, if natural intelligence is truly intent on forging an accurate reflection of itself, then it must, by necessity, employ a means to do so. The tree of life *is* such a means, its conscious aspect being the latest fulfillment of such a cosmic imperative.

THE NATURE OF CONSCIOUSNESS

Even if we grant that consciousness represents a sense-making evolutionary attractor in which natural intelligence comes to be reflected in a stunningly advanced way, mystery still surrounds the "stuff" of which consciousness is made. If we could comprehend of what "mindstuff" consists, this might shed some more light on its speculative function within the persistently constructive schemes of natural intelligence. Or at least we might be in a position to see more clearly how consciousness stands in its relationship to natural intelligence.

As an attractor, consciousness really is extraordinary and stands on a completely different level from other attractors such as the formation of nervous systems or metabolic pathways. Although such attractors represent extremely exquisite embodiments of organic intelligence, one suspects that consciousness represents some truly novel fulfillment of the natural order of the Universe. If the Universe is seen as a jigsawlike system whose components naturally self-organize into coherent patterns, then the emergence of mind within this cosmic system appears to represent a kind of dynamic *meta*-pattern. This is because for mind to arise there must perforce be a number of already extant patterns in place, in this case stable biological patterns such as advanced polyneuronal nervous systems. As far as we know, it is only when a biological system has evolved a certain threshold of polyneuronal complexity that conscious minds can arise.

Enduring patterns of consciousness therefore exist atop enduring patterns of polyneuronal bio-logic. Or, to put it another way, architectures of mind arise from architectures of polyneuronal circuitry. In this respect, mind represents an emergent higher-level aspect of the self-assembling jigsawlike Universe. In turn this seems to imply that consciousness is indeed special in some way, not just because it is so fundamental to our being, but because it can potentially reflect Nature to a strikingly clear degree. This latter realization gives good credence to the notion that consciousness represents natural intelligence in some sort of *tertiary expression* (recall that in chapter 5 I spoke of natural intelligence in terms of its primary expression as the laws of Nature and its secondary expression as the physical effects of those laws upon gross matter). What makes this tertiary expression so intriguing, of course, is precisely its ability to connect back to its source. In an act of comprehension, consciousness can quite literally connect itself to the very source of its own arising. Again, this would appear to be a kind of reflective process in which natural intelligence is mirrored by the human cortex.

MINDSTUFF

So what on earth is consciousness made from exactly? Having advanced the idea that mind is a dynamic architecture built upon highly evolved polyneuronal architecture, this still begs the question as to what stuff it is made of. Consciousness is obviously not made from fiberglass or aluminum, nor is it made of blood and bone. Conscious minds may well be inextricably linked to physical cells—which is what the neurons in our brains are—but these neurons would appear to act principally as an infrastructure, in the same way that the physical ink on the pages of this book is the infrastructure upon which the meaning of the text rests. So it is still unclear as to what consciousness *is*. While it may be a means through which natural intelligence reflects itself, it is still mysterious in terms of its substance.

Despite the difficulties in understanding the nature of mindstuff, there seems to be a general opinion among neuroscientists that consciousness is bound up with information processing and that it is something that the human cortex *generates*. Intuitively plausible, this is at least a foot in the door to an otherwise slippery domain of inquiry. In other words, the human cortex processes huge amounts of information and somewhere along the line conscious experience emerges. But it might well be the case that the human cortex *conveys* consciousness rather than generating it. This is not to say that each and every thought a person has is "out there" and is somehow picked up and conveyed by his or her brain alone, but rather that the "informational stuff" of which consciousness is made is, and always was, "out there." This implies that the human cortex serves to *focus* consciousness into being, that the potential for consciousness has always been latent within Nature and requires a highly evolved form of polyneuronal circuitry to "tap in to it," so to speak. So I am using the term "convey" in a rather loose way to help bolster the idea that consciousness is a potential that is focused into existence.

As an analogy, consider light and the evolution of systems that can focus light. But let us consider not biological systems this time but man-made systems. Here I refer to the development of photography, videography, and other techniques that serve to capture images from light. The first still cameras were of the pinhole type. They simply let in light through a small hole and captured the light information on a light-sensitive film. By today's standards, these earliest forms of photography are poor. Contrast this with modern digital video camcorders, which can capture temporal sequences of light information (approximately twenty-five quality stills per second). At the time of writing, high-definition digital camcorders saturate the market, having replaced regular digital camcorders and still older digital/analog hybrid Hi8 camcorders. With these compact portable machines, it is possible to record broadcast-quality images. Now 3-D video cameras are also coming on the market, and it is safe to assume that newly contrived formats like "ultra definition" or whatever will eventually follow.

Like many people, I own a digital camcorder. It is impressive that such a compact and lightweight device can generate high-quality HD film footage. What is amazing is to focus upon something, record, and then in no more than a few seconds be able to replay and review the footage on a small viewer screen. What one sees is an exact copy of what was previously filmed. A unique temporal sequence of light information (i.e., the impact of zillions of photons traveling at more than 180,000 miles per second) is faithfully recorded. Needless to say, the compact innards of these camcorders are highly complex, especially when you consider the millions of photons of light that must be accurately registered. Even the tapes, or hard drives, used to record upon are complex in terms of their surface chemistry, which has to convert light data into patterns of magnetically charged particles.

If we compare a digital camcorder with the older analog type and even the archaic pinhole camera, it is evident that what is changing over the years is not light itself but rather the technology for focusing and recording light. As technology improves and resolution is increased (more resolution equals more information from light), images become more and more refined, more and more accurate in their registering of light. In this sense, it is the sensibleness and intelligibility of light (i.e., the sheer richness of information within light) that drives camera technology. The constant evolution of video cameras is a technological response to the urge to accurately record the patterned information inherent in light. The more evolved or advanced the video camera is, the better able it is to faithfully record this information. And yet all this engineered complexity, all this convoluted design, actually pales in the face of the complexity that a camcorder is attempting to capture, *namely the ordered complexity of light itself.*

The important understanding to be gained here is that the most striking ingredient in today's modern camcorders and cameras is not their various ingenious electronic components or their zoom lenses or even the types of highly sensitive magnetic tape or hard drive that they utilize, but light itself, since it is light that contains the wealth of information we so love to capture and subsequently become enthralled by. The more evolved the camera, the better it is able to channel, focus, and record photic information. Of course, it is also the case that the patterns of light information must be processed by our brains before they make sense to us, but the fact remains that huge patterns of organized information are inherent within light itself.

Before tying all this in with consciousness, permit me to press home this point about light (after all, light is so very fascinating). Light clearly has an inordinate amount of temporal structure and temporal order to it and is richer in information than we can possibly conceive. For example, a camera so small that it can be perched on the head of a pin will take a *unique* photograph every time it is moved no more than a millimeter in *any* direction. Considering this, one can grasp the vast amount of information inherent in light as it continually ricochets around the environment. *All possible views exist, and they all pass through one another from every direction and at every conceivable point.* If we imagine a widely spaced row of a thousand people outside looking up at the stars, each person will have a unique perspective view, each will perceive a unique pattern of light. These unique patterns of light did not pop magically into existence when each person suddenly looked up. Each view, each streaming array of photic information, was always there waiting to be beheld. Move a little and a slightly different view comes into focus. When the first hot-air balloonists ascended high above Paris in 1783, their breathtaking view of the city from the air was likewise always there waiting to be seen. Similarly, we are rightly amazed by pictures of the Earth as seen from space, pictures that have been available only since the advent of space rockets in the 1950s, and that have been seen by the general public only since the late 1960s. And yet these spectacular views have always been out there, regardless of whether anyone, or any device, is there to register them. Although it is hard to grasp that *all* patterns of light exist in any one moment, reason decrees that this must indeed be the case. Thus, it is clear that light is ridiculously complex in terms of its pervasive nature and the massive amount of organized and sensible information it carries. And it was always a sensible property of the environment before eyes evolved to reflect its inherent sense.

Returning to the phenomenon of minds, *maybe the same can be said of the information of which consciousness is composed.* If, like light, consciousness can be said to consist of information (or complex patterns of energetic information), then it may be that what the human cortex does is to integrate and focus vast arrays of information into a "tight localized beam" of consciousness, much as a camcorder captures localized beams of light. Something along these lines must be true, since the human nervous system is exactly a kind of network, or system, of sensitive layers that absorb various forms of natural information. Our eyes sense and transmit light information. Our ears sense and transmit acoustical information (the molecular movement of air). Our noses also sense and transmit molecular information, although a different aspect from that which our ears sense. The same holds true for our taste buds. Our skin senses and transmits basic information about the shape, texture, and temperature of external objects. In each case, our sense organs—which are classic sense-making biological structures—absorb information about the world and then send this information into the mysterious depths of the human brain for processing. Somewhere along the line conscious experience results: I am feeling warmth, I am smelling vanilla, I am seeing the forest, and so forth.

The general idea, then, is that consciousness is the end result of the processing of information from many sources. Various forms of natural information, mostly external, are collated and sent to the brain, where an integrated conscious experience results. But rather than thinking of this as being the brain's method of generating consciousness, the entire process can be viewed as the brain's way of focusing or channeling consciousness into existence, in the same way that camcorders and cameras focus and channel light. And just as cameras and camcorders channel *unique perspectives,* or unique arrays of light according to their spatial position, so too can an individual human cortex be seen as channeling a unique array of information such that, over time, a unique psyche manifests. Since each human cortex represents a unique "angle of focus," as it were, and since each has a unique spatial position within a culture, the biosphere, and the entire cosmos, each will be imbued with individuality, in much the same way that all photographs or film clips will be unique.

RAYS OF CONSCIOUSNESS

If consciousness is a form of information, which the human cortex through its processing capacity brings into focus in the same way that cameras and camcorders focus light information, then maybe it would be appropriate to use the metaphor of a lens. Is the human cortex akin to a lens that focuses not light but consciousness? Are we all in possession of differently shaped lenses, thereby explaining our unique feelings of self? Are our individual perspectives on life and our memories thus synonymous with the personalized shape of our "cortical lenses," in the same way that uniquely sculpted glass lenses will focus slightly different patterns of light?

Whatever the veracity of these speculations, if the human cortex does indeed function like a lens—albeit made of billions of neurons as opposed to glass—that might imply that consciousness, or mind, exists outside of us (in some form) as part of the greater world. Individual minds could then be seen as "focused concentrations" of consciousness from a bigger cosmic source. In this respect, the informational content of the whole Universe may have mindlike properties in its own right, being a kind of distributed matrix of mind from which the cortex can channel a unique localized perspective—an individual "I," or individual mind, as it were.

As a radical metaphysical stance, the notion of a mindful Universe might well be true despite its fantastic air (after all, mind is assuredly a natural part of the Universe). Alternatively, our lens metaphor might simply imply that only the *potential* for consciousness exists outside of us, a potential that the lenslike human cortex can focus into existence. Certainly this latter idea would be more acceptable to most people, for, whatever you believe, it must be admitted that consciousness is a potential of the Universe that has become realized through the evolution of the human cortex (just as life itself was always a potential of the Universe). But whereas people of a merelyistic persuasion might just shrug their shoulders and state nonchalantly that Nature "just happened" to have a potential for eliciting conscious minds (as well as other remarkable phenomena), if we invoke the paradigm of natural intelligence, then consciousness emerges as not only a realized potential, but a potential that Nature was in some sense intent on implementing.

If consciousness, or mind, is a potential that becomes realized *only* through advanced polyneuronal systems, it implies that the Universe in itself is not mindful, or at least not mindful in a way that we can conceive of. Nature would still be intelligent at some fundamental level, but it would not be mindfully intelligent in the same way that we are (i.e., it would be an unconscious form of intelligence). It would also mean that the metaphor of a lens focusing light is not entirely adequate. The lens part might suffice, but not the light part. After all, we have established that light exists regardless of a visual system being in its vicinity. To improve our metaphor we require a *potential* aspect of light, some property or manifest phenomenon of light that arises only once a unique system that can channel light in a precise manner has arisen.

A laser beam immediately comes to mind. Lasers, such as those inside DVD players, emit laser beams, which consist of *coherent* light waves. Whereas regular light waves are "out of phase," a laser beam consists of concentrated light waves "in tune" with one another (of the same wavelength)—hence the term "coherent." Made up of coherent light waves, laser beams are powerful, and this explains why they are dangerous to look at directly.

As far as I know, lasers do not occur naturally in the Universe. Pulsars and quasars exist in the Universe, but astronomy textbooks do not list "lasars." For light to be coerced into a coherent laser beam, an elaborate laser device that can carefully manipulate light is needed. So even though laser beams are made of light, they emerge only when light is channeled and manipulated in a very specific way. Once this is technologically achieved—as it was in the latter half of the past century—then a novel potential of light is made manifest. Although the potential for laser light was always latent within Nature, it takes the sophisticated machinery of a laser system to generate an actual laser beam.

If we use a laser beam as our metaphor for consciousness, then it would mean that the human cortex acts more like a laser system than a lens, focusing a "beam of coherent information"—which we experience as consciousness—into being. Since the cortex is so stunningly complex, the idea that it works by integrating coherent patterns of information in the manner of a laser system is not too far-fetched. To be sure, it is quite an intriguing idea, and the more one accepts the existence of natural intelligence, the more compelling the idea becomes.

Imagine a designer race of sentient androids, each with a shiny shaved head inside of which resides an artificial electronic cortex and a stylishly designed logo like *Artificial Intelligence Inc.* etched thereon. We could view such an electronic cortex as akin to an information-organizing laser system, able to absorb, focus, and organize incredibly large amounts of information such that a concentrated laserlike beam of consciousness results (not shining anywhere, but emerging as a stable, self-organizing, informational construct). After all, that is exactly what a real cortex does and what an artificial cortex ought to do.

Picture such a race of conscious androids in your mind's eye. Picture their environment, the contextual surround in which they are completely enveloped, and the way in which this ambient surround constantly floods their artificial nervous systems with orderly and sensible information (via light, sound, etc.). Through this constant absorption of flowing information, ultimately channeled by the (artificial) cortex into a "final" coherent beam, a conscious awareness as to the meaning of themselves and their locale results, an impressive technological feat if ever there was one. It is in this sense that the actual human cortex is akin to an advanced (biological) machine able to focus a tight laserlike beam of consciousness into being. By being immersed in and processing a constant influx of highly ordered pattern-rich information, mind arises.

IMPLEMENTED POTENTIALS

Metaphors aside, we are still left with the question as to whether natural intelligence "set up" the potential for conscious minds. In other words, how does the Universe's immanent potential for brain-embodied consciousness relate to Nature's agenda? I have already suggested that consciousness, in particular consciousness of natural intelligence, is the evolved means through which a reflection of natural intelligence is more fully rendered. And I have also repeatedly pointed out that consciousness is assuredly a medium through which natural intelligence can know itself. Both notions are complementary, although the latter is perhaps simpler to entertain.

There are two forms of the conjecture that consciousness is the means through which natural intelligence can become more aware of itself. The first version simply states that the kind of consciousness with which we humans are endowed just happens to be able to apprehend natural intelligence. In other words, the fact that evolution has become conscious of itself, the fact that the Universe has become self-aware, the fact that we can divine intelligence in the laws of Nature and in the various manifestations of those laws, is just a curious evolutionary twist with no objective significance or purpose. Even if it were accepted beyond a reasonable doubt that conscious minds had to evolve, it still might not represent anything significant beyond being one particular instance of an evolutionary necessity. Indeed, a wholly conventional evolutionary biologist would probably say that consciousness, being a sense-making phenomenon, confers numerous survival advantages, thereby explaining its evolutionary emergence, and that it just so happens that an offshoot of this capacity is that we can learn all about the nature of the Universe and its fundamental laws.

Alternatively, the second version holds that there must be a deeper reason why natural intelligence has led to conscious minds. If consciousness is genuinely an attractor that eventually has to be reached according to natural law within most, if not all, trees of life, wherever they might exist, and if consciousness also represents a peak in the ongoing formation of a more accurate reflection of natural intelligence, then this implies that natural intelligence is somehow "intent" on knowing itself or exploring its own nature. Although we are breaching heretical territory once more, let's persevere on this course and see where it leads. After all, the last chapter is a customary opportunity to entertain a final round of radical speculations.

NATURAL LAWS AND NATURAL WILL

To bolster the idea that natural intelligence has intentional properties, we need to take another look at the laws of Nature (which I have repeatedly defined as natural intelligence in its primary expression). As we shall see, there is a close correspondence between laws and willful intent.

The most interesting feature of will is that once something is willed, once there is a will to do or realize an objective, things fall into place in the wake of that will. Take the wills left by people when they die. All the activity regarding the deceased's property and their finances is determined by the will they have left. Certain definite events transpire in accordance with that will, often creating a long and successive chain of occurrences that diffuse and spread themselves about in the larger environment. Or, if a man wills himself to exercise every day, then certain aspects of his behavior will be explicable only in terms of his will to exercise. Even simpler, if you will yourself to close your eyes, then your eyes will close—various nerve cells will fire and various muscles will behave according to your will. Will implies a *causal force* that sets specific events in motion. Will explains reason, purpose, and intention. It can also enforce order and cohesion. Moreover, as we know from our own experience, will is a definite property of the Universe, for we all have it, more or less. Like consciousness, will is as objectively real as are stars and planets.

Natural laws are especially indicative of will when we consider what these laws can achieve in terms of self-organizing processes. When it comes to fostering evolution to the point of conscious intelligence, natural laws most definitely appear significant and will-like. Whether the laws of Nature really are representative of will, albeit natural, is a moot point, which, as with most metaphysical speculations, cannot be proved or disproved. What can be said for sure is that laws are basically synonymous with rules, and that in the case of cultural laws the rules that compose them are designated by human intelligence. Human will/intelligence leads to cultural laws and cultural rules. These rules and laws then govern (or at least they are supposed to govern) human behavior and human transactions. Similarly, software programs are basically sets of rules and instructions composed by human intelligence that govern the ways in which a computer processes data.

In principle, wholly natural laws, such as the law of gravity, are similar to human laws and software laws, only in this case they govern the behavior of natural systems. Indeed, the laws of Nature quite literally *command* the cosmos to behave in an orderly fashion. Specific processes happen within Nature because the specific laws of Nature say so. The law of gravity determines why and how planets, for example, exhibit regular orbits around stars and why galaxies form. The law of gravity thus regulates gross matter throughout the Universe. Without this law there would be no stars or planets or spiral galaxies. And gravity can be mimicked inside a computer system by having the software written accordingly. Similarly, the laws of Nature ensure that all of the elements within the Universe bear consistent languagelike relations with one another, which allows for the emergence of the complex molecules of which life is made.

It is for these reasons that natural laws are akin to natural rules and regulations, able to weave and bind matter into specific sensible patterns. And since we know that cultural laws, rules, and regulations stem from human will and human intelligence, it is feasible that natural laws are bound up with a natural will of some kind. This is especially the case when we consider the ingenuity of natural laws in terms of their creative effects upon matter. And we should note here that Nature's laws are not quite as simple as some suggest. After all, the difference between the actual existence of a specific law and the non-existence of such a law is analogous to the difference between infinity and zero. Which means that the highly specific laws of Nature must somehow embody an inordinate amount of information. If something exists and it is informative (i.e., it can convey information), then it must itself hold information. In other words, because the laws of Nature serve to quite literally inform (and thereby shape) matter and energy in a precise way, they are in no way simple things, even though we might like to think of them as being simple.

Given that evolution has been fostered by natural laws to the point of conscious intelligence, natural laws therefore appear tremendously informative and curiously will-like. And given the inherent struggle, the concerted effort, that the tree of life must necessarily exhibit in its tenacious growth, this too is suggestive of a relentless force of will, or at least that life is actively animated by the orderly forces of Nature. Recalling that brief film in which three and a half billion years of evolution are sped up into but one minute, one can almost sense the biosphere flexing in response to the rest of the cosmos according to whose "magnetic-like" influence and order it is subject. But this is again to openly court heresy. According to the influential merelyism of Monod, for example, by intimating will within Nature we are in danger of succumbing to animism, of projecting human attributes upon the Universe.

While this might be the case, it is still true to say that the laws of Nature are will-like. Think about how science talks about the *forces* of Nature. The word "force" is interesting, as it suggests that certain events and processes are compelled to transpire. So once it is realized that the source of everything of interest in the Universe ultimately rests upon specific natural laws and that these laws force extremely interesting things to happen, then it not too far-fetched to interpret these natural forces as being inherently will-like. The naturally intelligent Universe can then be seen as a *deliberate* system.

Perhaps it is realizations of this sort that provide the impetus for some scientists to balk and thence embrace a multi-Universe paradigm. Such an ungracious move is a means of avoiding the implications of otherwise significant and ingenious natural laws. Personally, I am convinced of the reality of natural intelligence in the tree of life. In turn, I believe (albeit with a different kind of confidence) that the laws of Nature represent natural intelligence in its most primary form, and that the chief temporal consequence of this is the orderly evolution of the Universe, a kind of elaboration, or exploration, of this primary intelligence. If one accepts this premise, then Nature is not only intelligent but must in some way be intentional, or willful, or at least intention and intelligence are both aspects of Nature's self-organizing prowess.

The best bet regarding any "cosmic intention" is that it is bound up with self-reflection, in which case natural intelligence is in the business of reflecting itself (and thereby making sense of itself). Consciousness can then be seen as a potential means through which such a reflection is ultimately achieved. If consciousness can conceive of natural intelligence fully enough and act accordingly, then, according to this reasoning, consciousness is fulfilling its raison d'être. And who knows how many conscious trees of life are out there in the interstellar depths of space? If we are not alone—and the evidence strongly suggests that we are not—then consciousness might permeate the Universe. That would mean that a multitude of maximal reflections of natural intelligence might *already* have been fully rendered, a mind-boggling idea far too extensive to pursue in more detail here.

THE SELF-AWARE UNIVERSE

The idea that Nature is intent on making sense of itself and, by definition, intent on promoting a kind of self-knowledge has been alluded to by many thinkers, not necessarily of the overtly religious kind. In her book *Green Space, Green Time,* author Connie Barlow interviews (and cites) a number of scientists and philosophers who have an active interest in evolution and humanity's place within the cosmos. The following citation concerns answers given to the question of the putative function of our conscious species within the web of life. Some of these scientists we have briefly met, while others are new. I make no apologies for their poetical overtones:

> Edward O. Wilson suggested that we view our own species' contribution as that of "life become conscious of itself." Holmes Rolston proposes a slightly different self-image: "We are the love of life become conscious of itself." Half a century earlier, Julian Huxley urged us to consider ourselves as "evolution become conscious of itself." Thomas Berry and Brian Swimme envision our kind as the very universe become conscious of itself.
>
> Ten years after our species brought forth the first image of Earth from space, James Lovelock wrote, "The evolution of *Homo sapiens,* with his technological inventiveness and his increasingly subtle communications network, has vastly increased Gaia's range of perception.

She is now through us awake and aware of herself. She has seen the reflection of her fair face through the eyes of astronauts and the television cameras of orbiting spacecraft. Our sensations of wonder and pleasure, our capacity for conscious thought and speculation, our restless curiosity and drive are hers to share." Similarly, Joseph Campbell has said, "We are the consciousness of the Earth." . . . By all these standards, therefore, we are life, evolution, love of life, Gaia—even the cosmos—become conscious of themselves. All are *expanded self* images.[1]

Apt sentiments one and all. The gist is clear. Through mind, Nature becomes aware of its own fabulous structure and organization. When we observe the latest growth of the tree of life, spread as it is across the Earth's membranous surface, we are the biosphere become self-aware. Whenever we glance up at the night sky, we are the Universe become self-aware. When we observe a hummingbird hovering amid the heavy aroma of a nectar-rich spring flower, we are biological intelligence become aware of itself. Though we may not readily describe our experiences in such terms, in each of the above situations a sublime reflective process is most definitely at work.

THE CORTEX AS A MIRROR

If natural intelligence is indeed in some sense intent on knowing itself, or reflecting itself, the metaphor of a mirror also comes to mind. After all, the purpose of mirrors is to facilitate reflections. Although I am acutely aware that I am in danger of overburdening the reader with metaphors, I can think of no better way to make these admittedly lofty ideas more accessible. At any rate, one could imagine a human cortex opened out into a sheet. Often popular science books about the human brain ask the reader to imagine this. This is because the cortex is folded in upon itself in a convoluted manner but could, in theory, be opened out into a kind of surface roughly two-and-a-half-feet square. If all the cortices of every person now alive were to be unfolded in such a way and then stuck together, we would have a truly massive surface area. It is upon this huge cortical surface, polished by millions of years of evolutionary honing, that natural intelligence can, at least potentially, be more fully reflected. This is why the human cortex is indeed like a mirror. The reflection of natural intelligence is conveyed in the felt experience of natural intelligence. The human cortex once again resolves itself as the medium through which natural intelligence realizes a reflection of itself. As an attractor, consciousness is therefore the *necessary* stuff of which a more fully resolved reflection consists. Without drifting off into more esoteric domains pertaining to different states of consciousness, this is perhaps the most that can be said here regarding a putative function of consciousness in the light of the natural intelligence paradigm.

ETERNITY AND BEYOND

The keen reader will have gathered that metaphysics is a tricky, sticky area. In following the various implications of the natural intelligence paradigm, we will always be in danger of falling foul to fanciful speculation and unwarranted conclusions. However, one thing can be said for certain, and that is that regardless of what view you hold—whether you are a multi-Universe believer or an advocate of natural intelligence—when all is said and done an *uncaused* something or another *must* be invoked. And not simply uncaused, but eternal. For if we play the old multi-Universe game and assume that an infinite series of Universes exist, then we must admit that they have existed for eternity. They were, are, and always will be there—albeit without good reason. Similarly, if we invoke natural intelligence, then it must always have existed *in some form or another.* Even if, as some bright sparks have suggested, the almighty big bang happened due to some ostensibly random and bizarre quantum fluctuation, this still means that something, some state, some principle (mathematical or otherwise), always existed. And even if there was no big bang at all but only a "big rebound" from some previous contraction of the Universe, it still entails some eternally existing process or thing. It is misguided and facile to assert that the Universe sprang out of nothing. Whatever way you look at it, there was, at the very least, always a *potential* of some kind. And a potential is a real objective something, especially if it can lead to a whole Universe. Being a really existing thing, a potential cannot arise out of nothing but arises only by way of another really existing thing or another potential. In other words, you only get something from something else—this is an absolute logical necessity. It is unreasonable in the extreme to suggest otherwise.

The idea that something objectively real (no matter how abstract or exotic) has always existed, that something had no prior cause but was always in existence, is to conceive of an Ultimate Cause or First Cause according to which everything else happened. Which Ultimate Cause do we opt for? One that is totally devoid of meaning and intelligence or one that contains meaning and intelligence (and even mind) within itself? An agnostic will claim that nothing can be known about an Ultimate Cause (Darwin became an agnostic in later life, and a very weary one at that). Or perhaps an Ultimate Cause can be grasped only in a radically altered state of consciousness in which cause-and-effect cognition is transcended by some other kind of cognition.

For myself, I opt for an Ultimate Cause bound up with intelligence, with some principle that, if not truly indicative of purpose, is nonetheless in stark opposition to random mindlessness. If this is so, then the source of natural intelligence must somehow reside within itself. In that case, the big bang event, which spawned the entire Universe, might represent the creative outpouring of a powerful natural imperative. Moreover, fourteen billion years or so after this primal explosion, we are still being propelled upon its awesome cataclysmic impulse. The big bang (or big rebound) is still banging; creation is still happening. Though lowered in terms of energetic intensity, it is at this period of the creative explosion that consciousness has emerged, a kind of phase transition occurring in any sufficiently developed tree of life according to which the Universe can know its own nature and possibilities.

And so I leave the reader with a final thought: that a maximal reflection of natural intelligence *must* be conveyed by consciousness. Natural intelligence is borne anew within us, within the here and the now, within *this moment*. Einstein was arguably wrong when he implied that conscious human intelligence was but an "insignificant reflection" of the greater intelligence inherent within the Universe (as quoted in chapter 5). Consciousness, human or otherwise, is deeply significant. For consciousness is the means through which natural intelligence can truly know itself for what it is. In reorchestrating its informational substance through the evolution of life, by "turning itself inside out," as it were, natural intelligence is metamorphosed into the dimension of mind. Through sensing and feeling natural intelligence, by consciously awakening to the real situation in which we live and breathe, conscious experience *becomes* natural intelligence in its new form.

EPILOGUE

A Walk on the Wild Side

How are we to make sense of the spectacle of human history in which we are players one and all? Orchestrated out of pre-processed star dust, who or what wrote the human show, and why? If we embrace the paradigm of natural intelligence, at the very least we may see ourselves as fulfilling the dramatic cosmic imperative wherein natural intelligence effects a reflection of itself, a reflection of which mind is a crucial component. As to the ultimate consequences of this paradigm, these are anyone's guess. Indeed, the full implications of being alive and conscious within an intelligent self-organizing Universe are still to be explored. And what better place to dwell on such matters than in those areas of the planet where the biological face of natural intelligence is most apparent: namely, amid pristine wilderness. Let me therefore take you on a last wild and illuminating journey before we part company.

If, like me, you live in a busy city, there is much to be said for locating some wild tract of the biosphere and leisurely trekking through it with eyes wide open. Apart from offering us testimony to natural intelligence in its organic guise, the fresh air of wilderness country, the conspicuous absence of roads, cars, and human habitation, brings a welcome change of psychological impressions. If you leave a big city like London or New York and plunge yourself into the wilderness, the contrast can be joyfully overwhelming. The news, where "it's at" so to speak, now consists not of human affairs or the latest human fashions, but of everything happening around you, in the myriad forms of life that adorn such wild ecosystems. It's an innervating walk on the wild side in which to relax and celebrate the exquisite living substance of the tree of life. In a dazzling sequence of biologically wrought magic, the natural intelligence inherent in the wilderness seeps into the human mind, further reflecting itself, and thereby becoming known to itself. An ultra-smart circuit closes, bio-logic pulling off an alchemical trick of fantastic proportions. All is one and one is all.

The Lake District holds some of the most stunning and gorgeous areas of wilderness within the shores of Great Britain. William Wordsworth said as much, and I have no quarrel with his poetic sentiments. If one treks in the Lake District after having extricated oneself from the frenetic and polluted confines of a city like London, it takes a day or so to acclimatize to the new environment. The city is dragged along at first, as one's mind is still buzzing with city news, city friends, city gossip, and the afterimages of city life. But give it a day or so and the mind begins to calm itself. Gradually the sound of birdsong, the murmur of running streams, and the rustling of the wind as it caresses the trees begins to filter through into one's awareness, a gentle and unobtrusive chorus whose graceful tempo and smooth flow can only delight the senses. A keen observer will already sense that this is the live sound of natural intelligence, an orchestral performance welling up from the organic landscape whose players range from insects and birds to the undulations of water as it spills from the mountains and pursues its long and winding course to the ocean.

Two miles or so up above the village of Grasmere, where Wordsworth once held residence, wild and untamed ecosystems await the eager traveler. Overhead, solitary clouds drift across the azure depths of a broad sky, each cloud garnered, perhaps, from raindrops formed by ocean-dwelling algae. A roaring testimony to earlier downfalls, numerous waterfalls appear in the form of foaming white patches speckled across the mountains. These rare parts of the Earth's living hide bear an unmistakable aura. The inherent health of the various plants, their green vigor and sweet aroma, is refreshingly palpable. Left to its own intelligent devices, the landscape looks perfectly arranged, perfectly clean, perfectly manicured, a veritable efflorescence of life and vitality. Such wilderness scenes are no less than natural art, an open-air high-arts installation, a smartly organized three-dimensional tapestry, which, in contrast to human art, is of an entirely different caliber.

Sturdy juniper trees, stocky and almost bonsailike, stand majestic alongside tumultuous rivers strewn with smooth boulders of rock. Apart from fresh invisible oxygen, the junipers exude tangible organic vibrancy. Untainted by the encroachment of human civilization, such conifers appear astonishingly healthy, their allotted genes able to express themselves to concerted perfection. These are truly perfect specimens, impeccably realized forms of botanical intelligence. It is no wonder that the ancient peoples of this land, the Celts, worshipped Nature.

Turn about and magnificent oaks can be spied, dotted across the distant mountainsides on the other side of Grasmere. When one has learned to recognize the courtly oak from afar, one senses a uniquely wise expression of natural intelligence, a particular *technique* in the art of sensemaking slowly spreading itself across the landscape, colonizing, acorn by acorn, century by century. The oak and the juniper, the deciduous and the evergreen, two knacks, two autopoietic styles of natural intelligence, each flourishing imperceptibly, each branching, twisting, and writhing with metabolic life.

Dense families of fractal ferns adorn these paths less followed. When dried in the early autumn, the ferns are like delicate bits of the Mandelbrot set, crispy brown fractals scattered hither and thither as if some extravagant and carefree artist with a mathematical bent had once visited and moved hurriedly along after having cast his or her art. And then one begins to notice the myriad granite rocks, sitting firm like solidly enduring statements. Who can guess which massive glacier once carried these huge stones and eventually left them in their current repose? And long must they have been still, for each is covered with what at first appears to be ancient fluorescent yellow paint. The bright grainy paint appears everywhere, on all the rocks and all the boulders. An old marker perhaps, indicating that everywhere is the right place to be?

On closer inspection the luminescent paint reveals a complex inner structure. It is, in fact, a widespread lichen called *Rhizocarpon geographicum,* a deft symbiotic merger between a fungus and an alga. And there are other species of lichen here too; mottled ones, black ones, purple ones. Together, they form miniature ecosystems in love with the otherwise barren surface of granite. Any of these colorful lichen-smothered rocks could hold a place in the Louvre or the Tate Gallery. Each is festooned with tenacious and long-lived forms of symbiotic natural intelligence that began adhering itself to these rocks long before our great-grandparents were born.

Continuing our ascent we reach Easedale Tarn, a lake of crystal clear water ringed by steep and precipitous mountains. Though there is surprisingly little mammalian life here, a solitary hawk glides above the water, its screeches reverberating about the rocky inclines. A wakeful mind devoid of the usual habitual associations will see in the graceful flight of the hawk a particularly magnificent embodiment of natural intelligence. Not only can this integrated system of feather, muscle, and sinew take to the air with ease, it can immediately adapt to prevailing winds, steer a destination, and then land faultlessly, an enterprise of which a unique set of parameters and variables must prevail on each and every flight. This is a naturally intelligent capacity that would daunt the ablest human engineer. If we also consider that the hawk can locate food, unzip and re-zip the barbs of its feathers in order to keep them clean and in good working order, navigate, find a mate, build a nest, reproduce, raise young, and so on, then the full complement of natural intelligence that the bird embodies becomes emphatically pronounced.

And this is but the surface of the real truth. Deep within the hawk lies evidence of more sublime magic. The skull and bones of the bird, like our own bones, are made of calcium phosphate salts. Calcium is a natural hazard, a potential poison in its free ionic state, to which life had no choice but to adapt eons ago. By making prudent sense of calcium, by cunningly crafting the potentially lethal element into living architectures of bone, teeth, and shell, the tree of life thus overcame an otherwise life-threatening element. The memory of this naturally intelligent wile is therefore apparent, if we know how to look, in the hawk's delicately constructed skeleton, as it is in our own skeletal frame. So too with the mitochondria in the hawk's myriad cells. Like calcium, oxygen can kill. Contrary to what we might like to believe, the primeval production of oxygen by ancient life-forms was the biosphere's first environmental crisis. And yet through endosymbiotic bacterial marriages, the crisis was averted, oxygen being made sense of and put safely to work by mitochondria in the oxidative combustion of food. Who then is not astonished by the sheer breadth of natural intelligence evinced in a single hawk?

If the time is autumn, the green slopes around Easedale Tarn are home here and there to fungi. Spending most of the year entirely underground in the form of rootlike filamentous hyphae, these otherworldly organisms periodically break upward through the soil and morph parts of themselves into protruding mushrooms, which shed billions of microscopic spores, thereby fulfilling the obligation to reproduce. With their sticky surfaces, rubbery texture, and characteristic shape, fungi are justified in having a unique kingdom unto themselves. The kind of organic intelligence they evince is original to say the least. Being unable to photosynthesize, fungi must perforce practice autopoiesis in other sly ways. They can break down and ingest the nutrients held firm within fruit, bark, hair, horn, keratin, insect exoskeletons, and feathers. We do not witness this, of course, as the fungal recycling industry takes place beyond the scope of our naked senses. But we can know of it. And when we are wise to the nifty abilities of fungi, they appear more wonderful than ever.

Even the grassy tufts of this mystical region warrant our inspection. Sure enough, it is grass, but it is a noticeably different species from the grass in our gardens. This is a wilder type of grass, and there are many kinds. Peering more closely, there appear to be all manner of tiny plants intermingled with the grass. Miniature flowers appear, minute leaves, and a network of roots given forth by different species and yet intimately entwined with one another. The result is an intricate web of natural intelligence draped like a green patchwork quilt upon the landscape; exuding oxygen, imbibing carbon dioxide, and, ultimately, making sense of the entire context of Nature in which it exists. More than this, the dense green carpet with its roots and tendrils, shoots and stems, begins to look like futuristic *wiring,* a kind of massively distributed soft circuitry in which every living part, each and every cellulose fiber, is pulsating with naturally intelligent design and naturally intelligent functionality. The scene is actually more fantastic than science fiction, especially since it is real and evident to the senses. And how much more could we astonish ourselves if our eyes could be privy to the microcosmic dimensions of natural intelligence? For we should then perceive the vast colonies of bacteria that saturate our living mat and further enhance its symbiotic gist. Millions of species, millions of expressed genes, interlocked into a sublime flow of natural intelligence.

If one approaches a large granite outcrop, this gorgeously rich carpet of organic intelligence appears to be lapping against it, flowing imperceptibly, its border marking the flow's extent. One can even peel back this apparently still edge. A thin veneer of vegetation and soil comes away, a tangled conglomeration of hairy roots reticent to let go of the granite upon which it has alighted and from which it draws mineral sustenance. Return here in a decade and this vegetal surface, this gradual fluidic movement of natural intelligence, will be more apparent, for by then the edge of the green organic quilt will have ventured farther across the rock face. Given enough time, the mat will have completely covered the outcrop like some self-generating, self-repairing skin. Only the lay of the land reveals the rocks and peaks already enveloped below. An escalating tide of DNA-encoded natural intelligence sweeping across the Earth's surface, sustaining itself, defining itself, and becoming ever more coherent and interwoven. And ultimately we, like all other species, emerged from just such a tidal network.

In this luscious land we may even aspire to entomology. Insects abound among the unfettered undergrowth. Unfamiliar beetles the likes of which we have never seen in the city scurry this way and that, their green and blue metallic shells glinting in the sunlight. Mites, midges, and ants storm across tangled networks of roots. Large black beetles, looking like brilliantly armored tanks and undoubtedly the envy of any honest AI scientist, storm through the miniature jungle, oblivious to our investigative prods. Only in contrast to the designs of robotics engineers does the exceptionally advanced nature of these beetles really hit home.

Upturning a stone, an ant's nest beneath is suddenly galvanized into defensive action. In an event not likely to occur more than once in a millennium, worker ants are nonetheless immediately ready to react and begin to ferry pale white eggs belowground. Soldiers with oversized fighting mandibles amass and fervently seek for the source of the apocalyptic intrusion. The spectacle is one of orchestrated intelligence, a seething pool of intelligent behavior and intelligent structure, self-sensitive and ever vigilant to its surround.

Lifting another small rock reveals a small spider, dangling like a high-wire circus performer from a delicate strand of silk emanating from its abdomen. The spider momentarily spins about, its legs desperately knitting silk. It falls to the ground and, in alarm, rolls itself up into a tight fetal ball. This is novel behavior unlikely to be witnessed in the spider species that have adapted themselves to the dusty recesses of our homes. This is a defense pose honed in the wild, a neat trick to outwit a potential attacker. It lies there feigning death, a small innocuous bundle geared up to protect its unique allotment of autopoietic wisdom.

Venturing off the path, which winds its way up toward the distant mist-shrouded peak of Sergeant Man, the ground becomes wet and boggy. A different set of species resides here, perfectly adapted to the change in soil composition. Most noticeably we find dozens of species of carnivorous sundews, their sticky leaf tendrils waiting patiently to trap any insects who chance upon them. A quiet form of ambush, passive-aggressive as we might say, a fascinating drama played out between insect and plant. In the small pools of water that form here and there among the colony of sundews, we may also witness pond skaters, long-legged insects who wield a blatant miracle. Having evolved millions of years before the advent of Christ, these insects have learned to make sense of surface tension and can literally walk on water. It seems that natural intelligence knows no bounds in its sense-making capabilities.

If we choose to discretely camp out in a wild elevated arena such as this, new facets of natural intelligence become apparent in the dark hours of the night. In lighting a campfire, for example, we re-create the radiance of the sun (the reader must forgive such a questionable act, but not owning and driving a car, this is an example of my personal quota of pollution, pollution of one sort or another being an inevitable ingredient of life). A piece of dead juniper wood is effectively a long-lasting storage medium for condensed sunlight. By catalyzing an alchemical and endlessly alluring reaction between atmospheric oxygen and the carbon held firm within the wood, enchanting firelight and cozy heat are emitted, an echo of the sun whose radiation was originally channeled and packaged into the wood. The warmed human organism thanks juniper; juniper thanks the sun; the sun thanks Nature's fundamental laws. Consciousness basks in knowledge of the entire loop. An exquisite circuit is closed.

At night and from afar, the distant village of Grasmere looks like an attractive electronic machine nestled between the mountains. The implications of such a sight are so pertinent that we can be forgiven for turning our attention away from both the fire and the star-spangled Milky Way straddling the night sky above. Internally lit houses, streetlights, and glaring car headlamps moving in perfectly straight lines betray the presence of a technological species. One can reflect on the source of the energy that fuels the town. Linked by thick cables to a power grid in some nearby city, the town's energy supply is ultimately fossil fuel—coal and oil. And from whence come these fossil fuels, these instigators of light and electricity? The answer is the past work of bio-logic, the past endeavors of natural intelligence whose remnants changed over the millennia into buried deposits of coal and oil. By running on coal and oil, human culture has literally plugged itself in to the past evolutionary handiwork of natural intelligence. No work is lost. The collective auto-poietic efforts of times and species past are dug up and rechanneled into the service of humankind. The charged biosphere once again reorganizes its body in order to stimulate growth and potential progress, in this case the evolution of human culture.

And herein lies the great dilemma according to which the tone of this epilogue must now abruptly change. For while we strive to exploit the biosphere for our own ends in the name of progress, we have completely failed to concede natural intelligence, the very mainspring of our prosperous existence.

There are any number of ways in which the paradigm of natural intelligence can, and by rights ought to, change our global perspectives and our cultural value systems. This is the hope of any advocate of natural intelligence. Consider biotechnology, probably to be the single most lucrative global industry of the third millennium. With genetic engineering, genetically modified foodstuffs, cloning, and genetic therapy, the human race is seizing control of a power that, until now, was wielded solely by Nature. Apart from the difficult ethical and moral issues that the biotechnology industry raises, its chief fault is its complete failure to acknowledge natural intelligence. Biotechnology is concerned with manipulating those aspects of natural intelligence that are assuredly the most complex and the smartest. Genes are even patented in this furious industrial endeavor. Indeed, there has already been a race to sequence and thence patent the genome of rice, one of the world's most important crops. How can we have the audacity to patent natural intelligence and vend its legacy when the copyright is not ours?

When human scientists tinker with DNA primarily at the behest of profit and global markets, and often without a full knowledge as to the long-term effects of their manipulations, they are attempting to harness the essence of Nature's most advanced technology without having fully addressed the nature and source of this natural technology. The situation is a bit like us stumbling across the sophisticated computer technology of an extremely advanced alien race and then, through reverse engineering and various testing procedures, exploiting that technology, patenting it for our own ends, without even bothering to look into the full meaning and significance of the alien intelligence that produced that technology in the first place.

It is not that biotechnology is inherently wrong, only that the conceptual context according to which the biotechnology industry proceeds is devoid of allusions to natural intelligence, and therefore the direction of the industry will be misguided. Only if we are fully wise to natural intelligence can we know how best to proceed with our growing knowledge of the biological hardware and genetic software created by it. How can the human race possibly prosper if the biotechnological revolution is driven by a misguided paradigm of life? How can we know the right way to live and the right way to relate to the rest of the biosphere when our scientific paradigms have no room for natural intelligence, and we ruthlessly exploit the rich fruits of natural intelligence without contemplating the ultimate nature and origin of those fruits?

The paradigm of natural intelligence and its various implications should arguably be the yardstick by which we measure and judge all our global behavior. If we truly wish to progress as a species, the contextual environment in which our emergence was forged and that sustains us must surely be taken into account. We undoubtedly need to learn our true place in the scheme of things, to realize that we are not the sole arbiters of intelligence. Only if we learn from the intelligence embodied throughout the biosphere can we hope to steer culture upon the most sensible course. As we have repeatedly seen, it is sensible behavior alone that, in the long run, is preserved by Nature. This is as much true for the behavior of the human race as it is for the behavior of bio-logic. However, as yet, natural intelligence is not debated. Scientific journals, whether detailing genes or ecosystems, are devoid of references to natural intelligence. Human intelligence and human ingenuity, whether applied to genetic engineering or technological innovation, are what we celebrate, fund, and promote. By failing to recognize natural intelligence we are pursuing a blinkered course, oblivious to the great wisdom of Nature evinced all around us.

We can see in the current environmental crisis a further reason to reappraise our conceptual attitude toward life. As of 2011, the consensus opinion is that about a hundred species of organisms are made extinct each day due to land-use change and pollution. Considering that each of these lost species represents a unique concrescence of natural intelligence, a unique architecture of bio-logic borne of millions of years of evolutionary engineering, their loss to the world is a tragedy. If this loss pertained to works of art by Da Vinci or Michelangelo, the human race would be up in arms. And yet for all their glory and hubris, such human art pales in the face of the autopoietic art of natural intelligence. We, the conveyors and promulgators of human intelligence, are literally ransacking Eden. Our earthly garden, grown and tended by natural intelligence and able to provide for all our essential needs, is being wantonly decimated. Our tacit interpretations of life on Earth must surely figure in this disturbing equation.

The markedness of our conceptual shortfalling is apparent elsewhere. At the current time, figures show that the most noble of feline beasts, the tiger, is now down to a global population of little more than 2,500 mature breeding individuals. Why? Because of loss of habitat, along with the insane demand for tiger-based fetish medicine and tiger skins. The decline of the mighty tiger is thus a wretched symptom of man's greed, of man's insatiable lust to bend the biosphere toward himself alone, to bleed the biosphere dry of its most astonishing wonders. What do we tell our grandchildren? That the extinct tiger and thousands of other casualties were just organisms? Just gene machines? Just elaborate vehicles in the art of reproduction that can be dispensed of with little fear of responsibility? Or are such extinctions to be seen as part and parcel of man's ongoing *insult* to natural intelligence?

Insults to natural intelligence. Harsh words perhaps, but this is the truth of the matter. Each and every time the human race needlessly causes the extinction of an organism, each and every time we bulldoze a biodiverse ecosystem into submission solely in the name of economic growth, we are insulting natural intelligence. Consider something like the social prestige for owning mahogany furniture. By cutting down the aged and endangered trees of this species and processing them into expensive furnishings, we like to think we have produced distinguished art. Indeed, aficionados will drool over such furniture. And yet surely the true elegance associated with mahogany is more clearly evident in the tree while it lives. Mahogany is first and foremost an organism, not furniture. As a particular long-lived botanical embodiment of natural intelligence, the tree in its natural surround is far more worthy of our admiration than are the cultivated artifacts we draw from its ignoble death. As a living species of tree, the mahogany boasts all manner of ingenious properties, its unique legacy of DNA able, among other talents, to eat light and galvanize complex networks of metabolism. Plants are, in their own sedentary way, equally as complex and equally as judicious in their behavior as are animals—a fact Darwin himself was glad to help promote through his lifelong studies of natural intelligence in its botanical expression. Nonhuman organisms and wild ecosystems are not simply props or furnishings for the betterment of humanity. Rather they are embodiments of natural intelligence worthy of preservation and respect. One can learn wisdom from all organisms. Each is instructive in one way or another.

Insults to natural intelligence riddle human behavior. A person who has consumed alcohol to the point of vomiting is a particularly pertinent, albeit unsavory, example, capturing a more widespread principle in which natural intelligence is pressurized unnecessarily. Despite the earnest attempt to obliterate consciousness, heavy drinking forces the body into adept action. It is no easy biological feat to evacuate the stomach by means of sequential muscular contraction. Seeing such a person—and we have all seen one—reveals a strange juxtaposition. For here we have a stupid mind in possession of a wise organism. No matter how much poison is forcibly imbibed, the body will always react in an intelligent way. Indeed, the immense wisdom of the human organism bears no grudge no matter how and to what extent we abuse and mistreat it.

The same might well be true of the entire biosphere. With the worrying atmospheric increase of the greenhouse gas carbon dioxide caused by industrialization (along with changing solar activity), some scientists claim that the Earth is even now attempting to deal with this sudden imbalance. The contention is that the carbon within the excess CO_2 is being deposited, or "soaked up," by way of denser-than-normal wood growth in certain parts of the biosphere. As with the body's autonomic nervous system, the cybernetic wisdom inherent in the biosphere may yet prevent, or at least slow, ecological disaster. And this might be despite the fact that we are oblivious to such biospherical prudence.

And what of human war? How does the process of reciprocal mass destruction appear in the glaring light of natural intelligence? We often see casualties of war on television. It is not uncommon to see wounded civilians being treated in hospitals. Doctors will be seen desperately trying to stem the flow of blood from a bullet hole. A physician will know exactly what the victim's body is doing. The immune system will be in zealous overdrive. Operating on an unseen microscopic level, red blood cells and blood plasma will rush to the gaping wound to deliver their healing powers. But such a precisely effected wound is so traumatic, so insulting to human bio-logic, that the victim may have little chance of survival. Even the diverse and age-old natural intelligence embodied in the human organism cannot hope to cure the kind of savage injury inflicted by a high-velocity metal bullet. Guns, bombs, mines, and bullets are designed precisely to insult natural intelligence, and here lies their essential absurdity, for what is insulted is not even widely acknowledged. A technological war machine, crafted by human intelligence, is elevated to the peak of power and enforced respect, whereas the natural intelligence it serves to deliberately destroy remains unsung.

A similar absurdity occurs when we watch news footage of armies and ground troops storming across woods and wild valleys, embroiled in some war or another. Our focus is upon the soldiers caught on film. Intrigued, we watch them dig in behind some makeshift trench. Snipers take aim. Rockets are fired. Explosions erupt in the distance. Man-made flags, man-made borders, man-made racial concerns, man-made traditions of faith, and man-made political beliefs are invariably invoked to account for the eye-popping insanity. Who is likely to watch these dramatic scenes and deliberately shift their attention to the nonhuman elements? The biodiverse plant species stomped upon and crushed by boot and tank alike seem somehow beside the point. The lofty pine trees towering behind the troops are surely an incidental backdrop, the lichen and moss on their bark but a fleeting flash of color. The pollen-dusted symbiotic bee hovering about a soldier's rifle is likewise an aside, incidental to the main plot. A diminutive bird frightened into evasive flight by the army's intrusion into its territory is no more than an insignificant speck. As for the violent explosions, it is anybody's guess as to what amount of organically woven natural intelligence is torn asunder or choked to death.

Consciousness flows where attention goes. We are always at liberty to change the way we look at life. Natural intelligence and all its manifest wonders are here, there, and everywhere if we choose to become sensitive to them. As a paradigm, natural intelligence places things in proper perspective. By acknowledging natural intelligence, human intelligence is put in its rightful place, namely as an offshoot of an intelligent tree of life, moreover an offshoot that can potentially embody a consciously realized reflection of natural intelligence. By doing so, by aiding natural intelligence in its reflectional imperative, our conscious experience might yet expand, might yet extricate itself from the grim confines of history and, looking forward instead, serve to realize some new potential undreamed of in conventional philosophies.

The hope, then, the optimism to be seized and cultivated, is that despite our flaws, despite our greed and our destructive streak, the future of our conscious species may yet be splendid. Certainly we have the resources and the means. Our naturally intelligent biosphere is wired up. For something. Perhaps the information revolution, embodied as it is in networked computer systems spanning the globe, will aid, by virtue of a globally shared experience of some kind, the completion of an optimal reflection of natural intelligence. Which is to say that if natural intelligence is intent on knowing its own nature and possibilities, the collective human psyche and its extended nervous system of telecommunicational technology might well be the instrument through which the intention is further realized.

Of course, it cannot be conclusively proved that we live in an intelligent Universe. But I have at least endeavored to provide the reader with good and compelling reasons for thinking this to be so. Some might be asking for more obvious proof, as if natural intelligence were somehow synonymous with some godlike agency able to initiate a personal dialogue. Well, science itself is this dialogue, and there resides its essential merit. If we wish to learn from natural intelligence, it behooves us to familiarize ourselves with the findings of science. Whether this means that we pay closer attention to scientific knowledge, or we reflect on why we breathe before our last breath is done (and thus recall symbiotic mitochondria), or we strive to appreciate the sum total of intelligence inherent in the integrated biosphere, such a change of attention will help furnish what might one day become a continuous awareness of natural intelligence. As a paradigm, natural intelligence can liberate and reanimate the human spirit. Everything and anything can change in the wake of its cultivation. Natural intelligence makes sense—not just as a useful way of explaining life, but in terms of the ingeniously sensible structures and behaviors wrought by life. Indeed, natural intelligence has always made sense and always will make sense. Either we harmonize our behavior with natural intelligence and extend its potential, or we eventually cease to be.

The heated debates, which still rage over the veracity of Darwin's theory of evolution, may well indicate that the human race is at a crossroads. The apparent dichotomy of religion and science over the truth about how we came to be reveals that a deeply important issue is at stake. The reader is urged to embrace the theory of evolution and all that science tells us about the nature of the tree of life. But embrace it with a caveat. This is that the views promoted by orthodox evolutionary science do not provide the definitive solution to life's mystery. Indeed, a careful "contextualist" approach to understanding the tree of life yields an exciting glimpse of natural intelligence whose purposeful presence heralds a massive paradigm shift and more.

But to really feel natural intelligence beyond a purely theoretical understanding, to really sense the intelligence suffusing the tree of life, might well necessitate the occasional immersion in wilderness and the countryside. By leaving behind the overwhelming influences of human intelligence so overtly apparent in our cities, towns, newspapers, and broadcast media, we are more able to attune ourselves to natural intelligence. By so doing we may yet divine the full bounty of Nature's wisdom. Indeed, if our species has an inner yearning to invoke and honor something bigger and better than ourselves, let natural intelligence, in all its manifestations, be the subject of such spiritual veneration. This, surely, is the least that we can do while we walk the good Earth.

⌒

The fire crackles. Resin held within the burning logs of juniper begins to ooze, spit, and emit plumes of aromatic smoke. Having observed the nighttime spectacle of Grasmere and dwelt upon the global technological patina to which it is wired, our gaze is drawn back to the flickering flames dancing amid the fire. In the wake of profound contemplation, human intelligence begins to intimate the omnipresence of natural intelligence. Pupil becomes aware of Master. The conscious hominid brain, borne of three and a half billion years of ceaseless evolutionary effort, grasps the magnificently orchestrated system in which it has arisen. If consciousness exists, then so, too, must natural intelligence exist. With this heartfelt realization, an enchanting circuit is once more closed.

NOTES

Chapter 1. Why Are We?

1. Goldsmith, *The Way*, 200.

Chapter 2. Life: A Great Organization

1. Capra, *The Web of Life*, 81.
2. Ibid., 213.
3. Margulis and Sagan, *What Is Life?* 23.
4. Dawkins, *River Out of Eden*, 133.

Chapter 3.
Evolution: Never Mind the Gonads, Here's the Real Agenda

1. Dawkins, *River Out of Eden*, 120.
2. Darwin, *The Origin of Species*, 217.
3. Dawkins, *Climbing Mount Improbable*, 151.
4. Darwin, *The Origin of Species*, 219.

Chapter 4. Binary Acorns: The Science of Artificial Life

1. Ray, cited in Langton, *Artificial Life*, 180.
2. Ibid., 184.
3. Ibid.
4. Ibid., 187.
5. Pennock, cited in Zimmer, "Testing Darwin," 28.

6. Levy, *Artificial Life,* 164.
7. Ibid., 164–65.
8. Kelly, *Out of Control,* 398.
9. Monod, *Chance and Necessity,* 110.

Chapter 5. Oscillating Paradigms

1. Dawkins, *Climbing Mount Improbable,* 4.
2. Ibid., 23.
3. Ibid., 60.
4. Dennett, *Darwin's Dangerous Idea,* 59.
5. Ibid., 315.
6. Ibid., 177.
7. Darwin, *The Life and Letters of Charles Darwin,* 312.
8. Einstein, *The World as I See It,* 28.
9. Ray, cited in Langton, *Artificial Life,* 198.
10. Davies, *The Mind of God,* 81.

Chapter 6. Ah, But Can Nature Pass an IQ Test?

1. Kurzweil, *The Age of Intelligent Machines,* 21.
2. Attenborough, *The Private Life of Plants,* 41–43.
3. Dawkins, *A Devil's Chaplain,* 103.
4. Margulis and Sagan, *Microcosmos,* 232.
5. Kelly, *Out of Control,* 212.
6. Ray, cited in Langton, *Artificial Life,* 200.
7. Dyer, cited in Langton, *Artificial Life,* 119–20.
8. Minsky, cited in Kelly, *Out of Control,* 466.

Chapter 7. Close Encounters of the NI Kind

1. Margulis and Sagan, *What Is Life?* 72.
2. Margulis and Sagan, *Microcosmos,* 119.
3. O'Toole, *Alien Empire,* 195.
4. Attenborough, *The Private Life of Plants,* 108–9.

Chapter 8. Symbiosis: Making Sense Together

1. Margulis and Sagan, *Microcosmos,* 118.
2. Ibid., 119.
3. Ibid., 120.
4. Ibid., 152.

Chapter 9.
Attractors and the Evolutionary Emergence of Mind

1. Duve, *Vital Dust,* 292.
2. Ibid., 301.
3. Kauffman, *At Home in the Universe,* 20.
4. Ibid., 64.
5. Gould, *Wonderful Life,* 320–21.
6. Ibid., 290.
7. Ibid., 291.
8. Ibid., 310–11.
9. Duve, *Vital Dust,* 297.
10. Gould, *Wonderful Life,* 310.

Chapter 10. The Conscious Tree of Life

1. Barlow, *Green Space, Green Time,* 269–70.

BIBLIOGRAPHY

Attenborough, Sir David. *The Private Life of Plants*. London: BBC Books, 1995.

Axelrod, Robert M. *The Evolution of Cooperation*. New York: Basic Books, 2006.

Barlow, Connie. *Green Space, Green Time: The Way of Science*. New York: Copernicus, 1997.

Barricelli, Nils. "The Intelligence Mechanisms behind Biological Evolution." *Scientia* 98 (1963): 176–80.

Benyus, Janine. *Biomimicry: Innovation Inspired by Nature*. New York: Morrow, 1997.

Capra, Fritjof. *The Web of Life: A New Synthesis of Mind and Matter*. London: Flamingo, 1997.

Darwin, Charles. *The Origin of Species*. London: Penguin, 1982 [orig. pub. 1859].

Darwin, Francis. *The Life and Letters of Charles Darwin,* volume 2. London: John Murray, 1887.

Davies, Paul. *The Mind of God: The Scientific Basis for a Rational World*. London: Simon and Schuster, 1992.

———. *The Goldilocks Enigma: Why Is the Universe Just Right for Life?* London: Allen Lane, 2006.

Dawkins, Richards. *River Out of Eden: A Darwinian View of Life*. London: Weidenfeld and Nicolson, 1995.

———. *Climbing Mount Improbable*. London: Viking, 1996.

———. *A Devil's Chaplain: Selected Essays by Richard Dawkins*. London: Weidenfeld and Nicolson, 2003.

———. *The God Delusion*. London: Bantom Press, 2006.

Dembski, William. *No Free Lunch: Why Specified Complexity Cannot Be Purchased without Intelligence*. Lanham, Md.: Rowman & Littlefield, 2002.

Dennett, Daniel. *Darwin's Dangerous Idea: Evolution and the Meanings of Life*. London: Allen Lane, 1995.

Duve, Christian de. *Vital Dust: Life as a Cosmic Imperative.* New York: Basic Books, 1995.

Einstein, Albert. *The World as I See It.* London: Bodley Head, 1941.

Eisner, Thomas, and Edward O. Wilson, eds. *The Insects: Readings from Scientific American.* San Francisco: W. H. Freeman, 1977.

Goldsmith, Edward. *The Way: An Ecological World-View.* Totnes: Themis Books, 1996.

• Gould, Stephen Jay. *Wonderful Life: The Burgess Shale and the Nature of History.* London: Hutchinson Radius, 1990.

Holldobler, Bert, and Edward O. Wilson. *Journey to the Ants.* London: Belknap Press, 1994.

Juniper, B. E., R. J. Robins, and D. M. Joel. *The Carnivorous Plants.* London: Academic Press, 1988.

Kauffman, Stuart. *At Home in the Universe: The Search for the Laws of Self-Organization and Complexity.* New York: Oxford University Press, 1995.

Kelly, Kevin. *Out of Control: The New Biology of Machines.* London: Fourth Estate, 1995.

Kurzweil, Raymond. *The Age of Intelligent Machines.* Cambridge, Mass.: MIT Press, 1990.

———. *The Age of Spiritual Machines.* London: Phoenix, 1999.

Langton, Christopher, ed. *Artificial Life: An Overview.* Cambridge, Mass.: MIT Press, 1997.

Levy, Steven. *Artificial Life: The Quest for a New Creation.* New York: Pantheon Books, 1992.

• Lovelock, James. *Gaia: The Practical Science of Planetary Medicine.* New York: Oxford University Press, 2000.

• Margulis, Lynn. *Symbiotic Planet: A New Look at Evolution.* New York: Basic Books, 1998.

Margulis, Lynn, Celeste A. Asikainen, and Wolfgang E. Krumbein, eds. *Chimeras and Consciousness: Evolution of the Sensory Self.* Cambridge, Mass.: MIT Press, 2011.

• Margulis, Lynn, and Dorion Sagan. *Microcosmos: Four Billion Years of Evolution from Our Microbial Ancestors.* New York: Summit Books, 1986.

———. *What Is Life?* London: Weidenfeld and Nicolson, 1995.

Monod, Jacques. *Chance and Necessity: An Essay on the Natural Philosophy of Modern Biology.* London: Collins, 1972.

O'Toole, Christopher. *Alien Empire: An Exploration of the Lives of Insects.* London: BBC Books, 1995.

Perry, Nicolette. *Symbiosis: Close Encounters of the Natural Kind*. Poole: Blandford, 1983.

Prance, G. T., and J. R. Arias. "A Study of the Floral Biology of *Victoria amazonica*." *Acta Amazonica* 5, no. 2 (1975): 109–39.

Sagan, Dorion, and Eric D. Schneider. *Into the Cool: Energy Flow, Thermodynamics, and Life*. Chicago: University of Chicago Press, 2005.

Slack, Adrian. *Insect Eaters*. Totnes: Alphabet and Image, 2006.

Wilson, E. O., and B. Hölldobler. *Journey to the Ants*. London: Belknap Press, 1994.

Zimmer, Carl. "Testing Darwin." *Discover Magazine* 26, no. 2 (2005): 28–35.

INDEX

Amazonian water lily, 181

animism, 258

ants, 39, 157–66, 270

artificial intelligence, 3–4, 22, 69, 91, 112, 131–35, 139–40, 142–47, 160, 207, 253. *See also* intelligence

artificial life, 68–71, 83, 98, 131, 212

assassin bug, 165–66

astronomy, 242–44, 253

atheists, 52

Attenborough, David, 128, 178

attractors, 218–20, 223–26, 229, 231, 238–39, 245–46, 255, 261

autocatalysis, 205, 221–23, 234, 238

autopoiesis, 35–37, 39–42, 46–47, 89, 115, 127, 156, 164, 221, 266, 268, 270, 272, 274

Axelrod, Robert, 197–98

bacteria, 23, 41, 78, 86–88, 126, 137, 139, 155–57, 187–91, 194–96, 198–200, 225, 233–34, 237, 268

Barlow, Connie, 259

barnacles, 78–79

bees, 143, 158–59, 168–172, 178–80, 184

Benyus, Janine, 141

big bang, 96, 261–62

bio-logic, 7, 17, 78, 89, 127, 146–47, 188, 192, 197, 202, 206, 219, 227, 232, 239, 244, 265, 271, 273–74, 276

able to make sense, 4–6, 8–9, 22, 66, 136, 138, 174, 185, 193, 218, 225, 238–239, 241, 245

of barnacles, 79

cortical, 212, 246

evolution of, 24, 66, 82, 100, 114, 137

intelligence of, 26, 81, 90–91, 109, 113, 136, 138–39, 149

mammalian, 81

wisdom of, 84, 90, 137

biomimicry, 141–42

biosphere, 5, 11, 24–25, 27, 37, 39, 41, 85, 119, 127, 155, 157, 187, 190, 200–202, 230, 236, 258, 260, 268, 272–74, 276–78

biotechnology, 7, 142, 272–73

bladderwort, 177

brain

chemistry of, 15

endosymbiosis and, 199–201

evolution of, 8, 16–17, 23, 66, 80, 105,